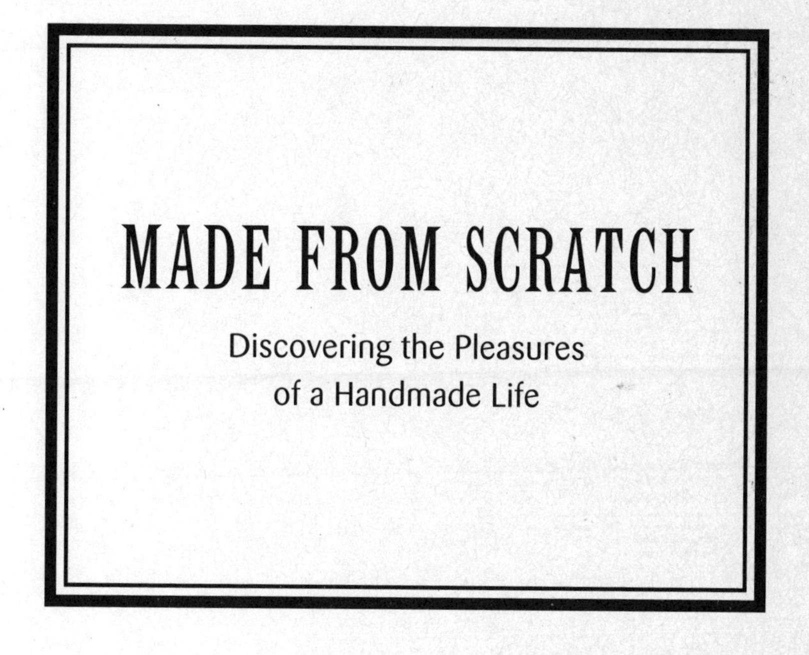

MADE FROM SCRATCH

Discovering the Pleasures of a Handmade Life

MADE FROM SCRATCH

Discovering the Pleasures of a Handmade Life

JENNA WOGINRICH

Storey Publishing

THANKS SO MUCH

Thanks so much to my parents, Pat and Jack, and my siblings, Kate and Johnny, who put up with me and my constant moving around the country all these years. Thank you, Diana and Bruce Carlin and everyone who was a part of Floating Leaf Farm, for your mentoring and, more importantly, friendship. Thank you to the Crawford family, who went above and beyond the call of duty as neighbors in Sandpoint. Thanks to the Tobias. Thank you, Carleen Madigan, the best editor in America, if only for the fact that she has a paint-by-number of Monticello over her desk. Also at Storey, thanks to Dan Williams, art director; Amy Greeman, publicity director; and Deborah Balmuth, editorial director. Thanks to Karen Kresge, my design professor, who pushed me so much to get my act together in college. Thanks to Kevin Boyle, Erin Griffiths, Nick Cooley, Heather Kerchner, Marie Hofer, Marjan Schelling, Erin Wert, Rikki Hamilton, Sean Sullivan, and Sara and Tim Mack, for taking the time to read through things with me and help out along the way with your advice and patience. Thanks to Raven Pray Bishop, who told me over the phone one night I should write this book and pushed me to get started. Thank you, Brian Campbell, for being the inspiration to all of us city-dwelling mountain folk, and thanks to Wayne Erbsen for writing those amazing books that taught me everything I know on the fiddle and banjo. Thank you, Kutztown University, for bringing design and amazing people into my life and leading me to Knoxville's door. And thanks to the Keystone, Gem, Green Mountain, and Volunteer states for everything you taught me. I owe you.

And, of course, a special thanks to Jazz and Annie,
the best roommates a girl could ever have.

FOR ANNA JUMBAR
(THIS IS ALL YOUR FAULT)

The mission of Storey Publishing is to serve our customers by publishing practical information that encourages personal independence in harmony with the environment.

Edited by Carleen Madigan
Art direction and book design by Dan O. Williams
Text production by Dan O. Williams
Cover illustration by © Christopher Silas Neal

Previously published in hardcover in 2008

Printed in the United States by Versa Press
10 9 8 7 6 5 4 3 2 1

Library of Congress Cataloging-in-Publication Data

Woginrich, Jenna.
Made from scratch / Jenna Woginrich.
p. cm.
ISBN 978-1-60342-532-2 (pbk : alk. paper)
1. Self-reliant living. 2. Country life. 3. Urban homesteading. I. Title.
GF78.W64 2009
640—dc22

2008034045

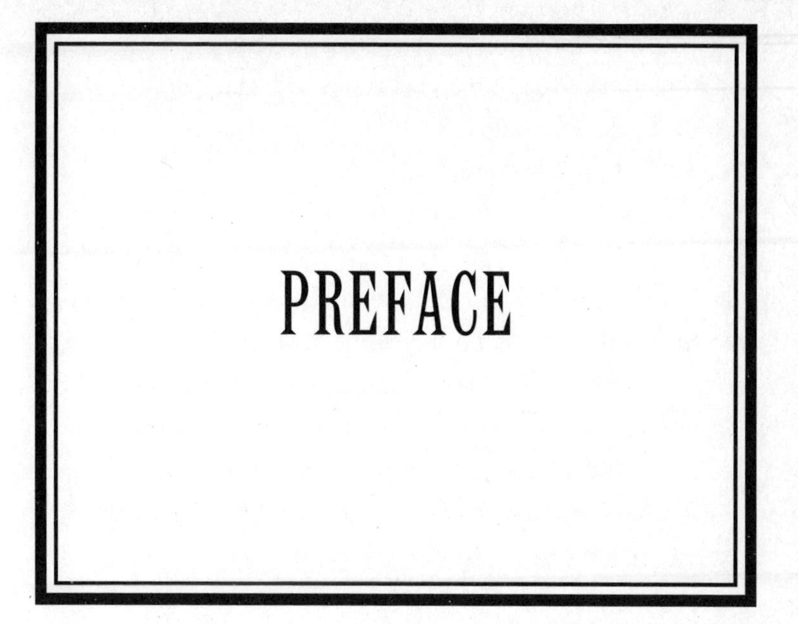

PREFACE

THE GENERAL OPINION IS, if you want a more self-reliant lifestyle, then you'd better run off to the country and buy a log cabin on five pretty acres. It should come complete with amazing views, a fixed-rate mortgage, and a porch swing — all of it miles away from the nearest shopping center or strip mall. Of course, we all know postcard property and an address aren't enough. You should also get a barn full of livestock, rolling fields of vegetables, and an old pickup truck with lots of rustic character. It also wouldn't hurt to have a shed stocked with every tool imaginable and a cellar stocked with rows of canned goods. Make sure there's a nice elderly couple next door to dispense advice. You know, someone to help you out when you need a cup of sugar, that kind of thing. (By the way, all of this should reside in some picturesque mountain valley where the word *reckon* is a regular part of the local vocabulary.) Heck, you could add all sorts of other requirements on top of that already tall order: renewable energy sources, a big green tractor, and maybe even a rumpled border collie sleeping on a feed sack in the sun.

Sounds heavenly doesn't it?

Sure does to me, but here's the catch. That general opinion leaves out a lot of eager people. Just because you won't be moving out of your apartment anytime soon doesn't mean you can't be more self-sufficient. Plenty of people in suburbs and sublets all over the country are replacing their pansies with peas and putting up henhouses where the doghouses used to be. Knitters are casting on in subways, and homebrewed wine is fermenting in your neighbor's basement. A revolution in self-sufficiency is riding the L train, and we saved you a seat.

I do many of the same things homesteaders do, but I don't own any land or even know how to start up a tractor. (Don't tell that to the guys down at the co-op — it'll destroy my street cred.) I also don't have a barn full of livestock or drive a truck (I drive a dented station wagon, but hey, it hauls its share of chicken feed). In addition to those shortcomings, I don't even work at home. I'm a nine-to-five corporate employee, and that's a position that doesn't lend itself to anyone's mental picture of *American Gothic.* (I guess if you replaced the farmers' pitchfork with a wireless mouse and gave him a pair of horn-rimmed glasses, you'd have a start.) Even so, if you're reading this, there's a good chance you work a day job, too. We all know the real world has real bills, and we can't quit our offices just yet.

What you *can* do, though, is change the way you live, no matter where you live. You can make better decisions every day; you can learn the skills that make for a more independent way of living. When you do, you'll start to feel more appreciation for those everyday tasks, because at the end of that day you're more in control of your life. That's really what this boils down to.

The work in this book isn't about playing farmer, it's about being more responsible for the tasks we've become numb to. We expect food to be waiting at markets and entertainment to be a few buttons away. When you start producing your own food, even the simplest plot of potatoes, your life regains some of the authenticity we've all forgotten about. When you sit back against a tree with a mandolin on your lap instead of lying on the couch with three hundred channels of instantly recordable distraction, you gain a little more from your downtime. You'll find yourself more humbled, satisfied, and grateful to have found a balance that simplifies your life with the skills of the people who came before you.

Point is, it feels good to get dirty, work hard, and slow down.

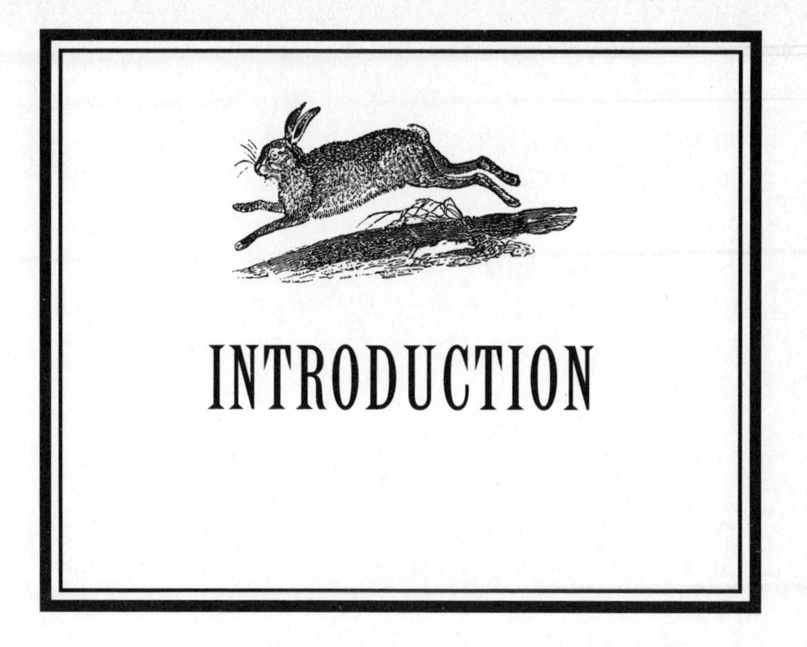

INTRODUCTION

LATE ONE NIGHT I was grinding coffee and listening to a radio show. There was nothing particularly interesting about this. Most nights I get the percolator ready for the next morning, and the radio is almost always on in the kitchen. But that night I realized something.

A hundred tiny efforts and decisions had converged right there on the countertop. The radio was crank powered, and the coffee grinder was an old hand-turner I got at an antique store. I was standing in the glow of my solar-powered lamp with the aid of some beeswax candles. Suddenly I realized that nothing I was doing required any outside electricity. I was seeing in the dark, grinding local beans, and listening to renewable energy–driven entertainment — and, as mundane as the situation was, it felt perfect in its order. Outside the kitchen, my trio of hens were cooing in their hutch, and snap-pea pods, hanging heavy on the vine, were climbing up my windowsill. The dogs sighed and stretched on the kitchen floor and the smell of just-crushed coffee beans wafted through the air, giving me a sense of profound comfort. I felt that if the world shut down, we'd just go on grinding and stretching and sighing till we retired to a warm bed. Maybe it was the candlelight or maybe it was the promise of fresh coffee in the morning, but in that moment I felt I'd accomplished more than anything I had ever achieved in my professional career.

MY FIRST STEP down the path toward self-sufficiency happened when I started learning more about how products get to us consumers. I was considering a vegetarian diet to get in better shape and feel healthier, and by reading a few basic books on vegetarianism, I learned about the mass production of meat in factory farms and all its related problems. The more I educated myself about how the meals I was eating got to my plate, the more disgusted and disappointed I became.

I also became much more appreciative of small farms. What started as an envy of picturesque, weathered red barns and gamboling herds of animals became a deep respect for the role of small farms in producing food humanely and safely. The more I read about all the small organic farmers who treated their meat animals humanely and didn't flood their planting fields with chemicals and pesticides, the less I could stomach buying those foam trays of meat and plastic bags of vegetables in the grocery store.

When you begin to comprehend how something as basic as how food gets to you, you begin thinking about how other items find their way to you, too: things like clothing, produce, and especially energy. The bloodshed and national security threats caused by depending on foreign oil were loud and clear on the daily news. The scary thing was that I was completely dependent on fossil fuels, and so was everyone I knew. My gas-heated apartment, my groceries from the supermarket, my station wagon parked outside — everything was part of the system. And if the system broke, I was going to be hungry, cold, and immobile.

So I threw my hands in the air. I was done with Walmart and Wonder Bread. I wanted something real. I wanted something basic. I wanted a lifestyle that was no longer a part of the problem, or at the very least was constantly striving to be less involved in it. I wanted a more sustainable life. Learning about homesteading — or the skills associated with it, anyhow — seemed like the solution I desperately craved. I decided to take the reins and start learning how to produce some of the food, and I used resources like the ones in Research, Son, every day.

There were obvious problems. I'm not exactly Gene Logsdon here. I had just spent four years in design school, learning where to put things on computer screens, and that doesn't help you when you're trying to bed down a chicken coop. Also, I didn't have a home to stead. I didn't even have a backyard. And the only skills I loosely possessed were simple knitting and soapmaking, which I did for fun, not as intentional parts of some self-reliant lifestyle. So I opted to engage in some simple research.

I PORED THROUGH BOOKS AND MAGAZINES. I haunted homesteader blogs and online forums. I did whatever I could to edge my way through the crack in the door. But finding a mentor who could teach me in person made all the difference. My first visit to a coworker's farm one Saturday in February turned an evening of conversation into an amazing friendship and a year of learning a more self-reliant way of life. It seemed ironic that I didn't meet Diana Carlin at the farmers' market, or even in the produce section of the grocery store, but at a giant corporation. Her cubicle was a few feet from my own at work. A couple of weeks after we met, she invited me over to take a tour of her family's homestead, about twenty minutes from the office, meet the animals, and get a personal introduction to chicken farming.

Diana's house was exactly what I imagined a real homestead would look like: a long, cedar-sided house with a chimney that puffed a wispy trail of wood smoke. It was surrounded by meandering homemade fences and was tucked into a spread of hills. A few pairs of cattle plodded around the front yard, grazing on the lawn (I wondered if the Carlins ever had to mow).

We spent the daylight hours meeting cows and collecting eggs from her hundreds of hens. After we tended to the animals, washed up, and ate a good meal with her family, we retired to the couches to talk shop. Maybe it was just my full stomach, but I felt comfortable. Usually when I meet new people I'm guarded and slightly nervous, but here I felt content. I sat back against a cowhide, which I was told once belonged to Ronald, one of their first farm-bred steers. If my vegetarian friends knew I was in an Idaho farmhouse snuggled up next to a blanket with a name, I think they'd be disgusted. But I'm a practical sort of vegetarian. I became one because of how meat gets to the table — the disregard for animal welfare and the assembly-line style of death was too much for me to get any enjoyment from a fast-food hamburger. But here at Floating Leaf Farm, everything was done the way it had been before industrialization became the norm. I respected that. I leaned back into Ronald (who was very warm, by the way — who knew cows were so woolly?) and accepted a glass of homemade wine.

Turns out Diana's husband, Bruce, unconditionally loves three things besides his wife: Italy, wine, and Italy. He's always been a connoisseur, and as a couple they've traveled all over the world visiting vineyards. Bruce's love of good wine has also driven him to make and bottle his own at home. They had everything from the fruitiest Rieslings to the darkest Super Tuscans, all bottled on the premises. Over the next few months, I heard stories about everywhere from Italian vineyards to garage bottling operations in the Washington backwoods, all told with equal excitement and devotion.

I learned more about beekeeping over three glasses of Bruce's wine than I could've learned reading through the entire B section of the encyclopedia. While we were chatting and sipping Syrah, I brought up the topic of honeybees, mentioning that I'd always wanted a hive. Something about them is so intriguing — you have no idea how much is going on inside those wooden stacks of comb-laden frames until you start really learning about it. Diana had a few hives and even had one propped up on the porch just outside the sliding glass doors of her living room. I began with casual questions, and what started as innocent small talk snowballed into a full-blown crash course in beekeeping.

Diana talked excitedly about queens, drones, workers, hive bodies, and nectar flow. She laughed loudly over the horror stories of failed hives and multiple neck stings. When she lifted up a screen and pointed out the parts of the comb, proudly displaying where the rich, thick honey clung, she made the process seem like a kind of religion. Her egg sales at the office had covered the cost of a honey extractor, she said.

I nodded, my eyes wide, and hugged my elbows. If it was cold out, I didn't mind. Thanks to the Carlin family and a few glasses of wine, I was plenty warm . . . and so inspired. The animals, the farmhouse, the happy family — Diana had accomplished everything I'd dreamed of. She was proof positive that a modern homesteader can have it all.

Through the long winter and into spring, Diana helped me get going with my livestock efforts. With her help I got a small flock of chickens, two long-haired Angora rabbits, and a hive of Italian honeybees to buzz through the garden. On summer nights at her place, there would be barbecues and campfires with music and friends. On calmer nights, I'd relax in a hammock on the back porch and watch what she called "Farm TV." It was more engrossing than a Ken Burns documentary and more entertaining than a good sitcom. I'd sway back and forth, watching the calves chase after roosters and ducks waddle about the creek. Agatha, the Carlins' gray cat, sauntered past the red barn, ignoring the chicks at her feet. Angus and Bella, their two dogs, loped along the back pasture. I was mesmerized. Every so often Diana would come out to check on me and look at the episode I was watching, and she'd say, "Oh, I've seen this one already. Damn reruns." I'd laugh, and she'd pour more red wine into my glass. Life rolled.

Diana and her family taught me everything from pounding fence posts to making and canning tomato sauce. It was the best type of mentorship a person could have. Even though we started off as strangers, I felt like I had become a part of their family.

. . .

AND SO my adventures began. The same mess of hope and fear lies at the beginning of any adventure, but just deciding to engage in the things that keep you alive might be the most hopeful and fearsome part of it all. It's rewarding in its simplicity — the garden, the egg, the music and friends, the new people and conversations on porches along the way. It's perfect, all of it, if you just let it be. If you can sit back and just take in the experiences, paying attention to every one, I promise you'll always come back to them. You'll lie awake at night thinking about the joys of holding your first baby chicks in your palm and the bliss of serving a salad from your own garden. There is something in these actions that fills you up.

I still dream that someday I can support myself without an office job, and maybe someday I will. Until then, I'll keep running my renter's homestead. I'll keep producing my own food, tending my small livestock, and canning my own sauce, because it makes me feel more in control of my day-to-day life in a way the cubicle never could. I've come to understand that what I do in my professional life is not as crucial as I had thought. When I realized that the heavy stuff, the *real* stuff, was back home on the farm and not at my desk, everything changed. After a few months on the farm, even the biggest crisis at work was just something to be dealt with calmly and rationally. Other employees would act like a deadline was a hurricane, but when you'd spent the morning deciding whether or not a rabbit with a broken spine would be put down, you couldn't really stress over PowerPoint presentations. Ironically, it was starting my own homestead that made me happier at work. Go figure.

I THINK THE REAL TRICK to finding that sense of satisfaction is to realize you don't need much to attain it. A window-box salad garden and a banjo hanging on the back of the door can be all the freedom you need. If it isn't everything you want for the future, let it be enough for tonight.

Don't look at your current situation as a hindrance to living the way you want, because living the way you want has nothing to do with how much land you have or how much you can afford to spend on a new house. It has to do with the way you choose to live every day and how content you are with what you have. If a few things on your plate every season come from the work of your own hands, you are creating food for your body, and that is enough. If the hat on your head was knitted with your own hands, you're providing warmth from string and that's enough. If you rode your bike to work, trained your dog to pack, or just baked a loaf of bread, let it be enough.

Accepting where you are today, and working toward what's ahead, is the best you can do. You can take the projects in this book as far as your chosen road will take you. Maybe your gardens and coops will outgrow mine, and before you know it you'll be trading in your Audi for a pickup. But the starting point is to take control of what you can and smile with how things are. Find your own happiness and dance with it.

CHICKENS

The most exciting backyard accessory since lawn darts

MY FIRST HANDS-ON EXPERIENCE with my own laying hens happened on a snowy March morning. Everything outside seemed heavy and wet, and the boughs of cedar trees along the road buckled under the uninvited weight of snow. Regardless of the miserable conifers, I was giddy. I was listening to bluegrass on the car stereo and singing along in the annoying way that makes other drivers stare at you at red lights. I didn't care — it was chicken day! I was driving south on Highway 95. My destination: a rural post office a few miles outside of town. I had been instructed to drive around back and knock on the steel door, where apparently some postal worker would hand me a box of fifty chirping day-old chicks. I couldn't decide whether I felt like a seven-year-old on Christmas morning or an agent sent on some kind of agrarian secret mission. After years of dreaming about farm animals, I was finally moments away from having my own.

Most hatcheries ship in the spring and require a minimum order of twenty-five birds, which seemed like too many for a beginner like me to take on. Fortunately, when my friend Diana was ordering her first batch of chicks for the year, she invited me to tack on my humble order. For the unbeatable price of $2.85 apiece, I purchased five Japanese Black Silkie Bantams. I chose Silkie Bantams because I'd read that they were gentle with people and also amazing mothers to their young. Some farmers have been known to take eggs from other breeds and slide them under Silkies to ensure decent parenting when the chicks hatch. Bantams are also smaller than the average, seven-pound chicken, some weighing as little as two pounds and standing half as tall. As someone who might be packing up her flock in cat carriers if she moved, I appreciated their compact size.

Besides their smaller stature and gentle demeanor, Silkies lay perfect little light brown eggs that are about half the size of the ones you'd buy at the store. Five hens meant I'd have well over a dozen quaint eggs a week — more than enough for a single-person household. As an added bonus, the hens look hilarious: pitch black through and through, with a clownlike poof of feathers on their heads (and they weren't anywhere near being the silliest-looking birds in the hatchery catalog, either). They looked nothing like the generic farm chicken, and that suited me just fine.

AFTER A FEW SONGS, I arrived at the post office. It was a small building you could've mistaken for a storage shed if you drove by too fast (which I did, and had to turn around). When I knocked on the back door, an old man with a basset hound greeted me and handed over the loudest cardboard box ever. I placed it in the backseat of my toasty Subaru and continued down the road to Diana's homestead — Floating Leaf Farm. A steep hill and some twists and turns later, I pulled into the driveway of the cedar-shingled, log cabin–style farmhouse. After all the snow we'd had recently, it would've passed for something on the label of a maple syrup bottle if it weren't for the year-old steer walking down the driveway to join the rest of its gang.

Diana, who reminds me of a mother hen herself, met me with a smile at the basement door, and I handed her the fifty-bird box. I followed her into the boiler room, where she'd already prepared a brood box for her new arrivals. The room was so warm, it felt like a natural incubator for the new kids. My birds wouldn't get such grand treatment until I took them home, so for the meantime I had brought a small cardboard box with cedar shavings and a hot-water bottle to keep them warm. My Silkies wouldn't be ready for another road trip, though, until they had settled in with some water and snacks after their long journey to the Pacific Northwest from the hatchery in New Mexico.

We set the box on a countertop, and Diana cut open the straps and slowly removed the lid while I tried to contain my excitement. What lay within were the most adorable contents of anything I'd ever gotten from a post office: fifty meandering, fluffy little chirping chicks. It looked like the hatchery had sectioned the aerated box into four main compartments to keep the squirming, peeping poofballs from piling up on top of each other. Half of them were the kind of fat yellow chicks I'd assumed all chickens look like at that stage (these were Cornish roasters for the family's yearly meat supply), and the other twenty-five looked more like confused little sparrows: tawny, black-capped brown birds with black eyes. These were Diana's laying hens, a breed called Barnevelder, which would grow up into handsome brown-egg layers. I scanned the carton for my little black babies, but they were nowhere to be seen. Diana noticed my concern and moved a handful of Barnevelders to the side. In a cowering huddle below them were the tiniest black birds I ever saw — I could hold all five in one hand.

We had to get them out of that shipping box, and soon. The brooder Diana had set up was her own invention: a small, foot-high fence of chicken wire draped with old quilts and blankets. A heat lamp hung through the blankets into the brooder, where there were food and water stations and a comfy layer of cedar shavings on slabs of cardboard to insulate the cold concrete floor. One by one, we picked up each chick and dunked her little beak in cool water before setting her beside the tiny watering font (an upturned Mason jar with a screw-on attachment that allows water to be released slowly). The chicks opened and closed their little beaks and drank the water, which they had just realized they desperately wanted.

"Make sure each bird gets a drink and knows where the fonts are before you move on to the next one," Diana instructed patiently. I was lucky to know someone who was so willing to teach me this stuff. Learning from books is one thing, but actually helping set up fifty chickens for life on a farm was a hands-on lesson I could never forget. I treated my girls the same way and gave them a chance to stretch their legs in the warm brooder before transporting them to their taxi.

. . .

WITHIN THE HOUR, I was driving back to my farmhouse with a still-chirping but much quieter backseat. When I got home, I rushed them inside to a larger cardboard box on the kitchen counter. A brooder light shone above it, and the thermometer I had taped to the side of the box read 90°F. Perfect. Inside was a little plastic font and a similar contraption that held tiny starter feed crumbles. I gently placed the babies into their new home. For a few weeks, this would be their world. I watched them mill about, pick at their feed, and drink from the font, until finally they all piled together and fell asleep.

This was it. I was a chicken farmer. I spent the next few hours in the kitchen, watching them as if they were being interviewed by Charlie Rose. I was completely enthralled with their journey over the past few days and with having so far pulled off my part of the deal — which was making sure they stayed alive and happy. I sat there staring at them, and as corny as it may sound, I felt a little bit like a parent. A parent with a very specific plan, though. All I could think about was what the summer would be like with these girls strutting around the backyard, laying eggs for me every day.

Well, I'm sorry to say those little birds never did get to strut around my yard, or any backyard, for that matter. My otherwise sweet Siberian huskies got into the brooder box four weeks later when I stupidly left it, unsupervised, at their eye level. In the ten minutes I was gone from the room, they killed all five of the chicks. I had the small consolation of knowing my Silkies had a swift death, but it still had me crying and calling Diana. Sadly, losing livestock — whether it be at the jaws of predators, to disease, or to natural aging — is a reality that even a hobby farmer with a painted-cottage coop needs to deal with.

After some discussion, we decided that I would try again with new chicks when she placed a meat-bird order later that month. This batch of Silkies I would raise in the garage. So that I could have some chicken-rearing experience before they arrived, Diana gave me three of her adult birds for the backyard coop I had already set up outside my kitchen window. Kind of like training wheels.

Enter Bertha the Buff Orpington, Astoria the Australorp, and Glowbug the Light Brahma. By cover of night, when Diana's two hundred chickens were sleeping on their roosts, we snuck in and kidnapped the three giant, heavyset hens. Apparently, transporting chickens at night is the only way to go. They seem to be calmest and most comfortable on a new roost when they wake up there. With little trouble, we settled the three girls into a carrier and put it in the Subaru. I drove them the twenty miles to my place, where we took them behind the house to their new coop — a raised hutch, lined with hay and stocked with a sturdy roost and laying box. After we settled the birds in, I closed the wire door and latched it. Within minutes, they were fast asleep.

Well, I didn't have much luck with those hens, either. Astoria took off for the woods as soon as I opened the door to the hutch and was never seen again. The other girls stuck around but just freeloaded. They didn't lay a single egg. I thought the purpose of having laying hens was, well, *laying.* But those biddies were more interested in pest control than production. All they did was cluck around the pine trees, eating bugs and snapping at flies. After a few weeks, I sheepishly approached Diana about a trade for some younger birds. I felt guilty, after all she'd done for me, but she just laughed. I guess when you have hundreds of chickens, you don't take individual character analysis personally.

Bertha and Glowbug went home to their old roost and three new gals took their place: a Red Star, a Buff Brahma, and a Welsummer. These girls didn't come with names, but it was pretty easy to hand them out the following morning. I woke up to such a clatter I thought they were being ripped apart by a coyote and ran outside in my pajamas. Inside the safely latched coop was the screaming Welsummer. She was ticked off to wake up in a smaller, less fancy space, and she wasn't going to take it quietly. I didn't know what to do. She didn't want her feed or water. She just wanted to carry on. I thought the neighbors were going to call the police, so I opened the coop and scooped her up in my arms like a kitten. She defecated all over me. Fine. I set her down to explore the yard. That didn't help at all. She just kept walking around screaming even louder. I named her Ann Coulter.

The other girls were quickly named as well. Veronica was the proper name for the quiet and cooing Brahma. A real sweetheart, that one. And Mary Todd Lincoln was the name for the crazy Red Star. These girls turned out to be champion layers, each of them laying a different color of egg. Mary Todd's were a classic warm brown, Ann's were a dark, speckled chocolate, and Veronica's were a classy cream. Sometimes I'd get the coveted trifecta: all three eggs waiting in the nest box when I came home from work.

Having chickens was so much fun and so rewarding, I wanted more. Soon I had the new Silkie chicks in the garage and my convent in the backyard, but I wanted to add some middle ground to the mix. With young chicks inside and my used birds with their own levels of personal mileage already established, I thought it would be wise to get a young laying-age hen to join the menagerie.

So, I ordered an eighteen-week-old hen from a hatchery. These adolescents are called "started pullets," and if you want eggs soon, they are the way to go. I named the new hen Mindy. She was a pretty red production layer like Mary Todd; in three weeks, Mindy started laying tiny pullet eggs, and a few weeks after that, she was churning out a whopper a day, some with two yolks inside!

That's one of the perks of having your own chickens: you get to see all the crazy egg mutants that aren't sold at the grocery store. There are giant double eggs that can't fit in regular cartons, and flat-sided eggs, and bumpy eggs that came out too fast. Some are dyed only halfway by the bird's vents, as if they'd been dipped in varnish. (Vents are chickens' all-purpose exit chutes. They lay eggs from vents, and other necessaries exit that way, too, if you catch my drift.)

By high summer, my little flock was producing up to eighteen eggs a week. They ran around the yard happily clucking; the Silkies, now old enough to join in, scratched around, tapping out Morse code to the big girls' more elegant prose. Friends who stayed over woke up to juvenile Bantam rooster crowing and could be introduced to the exact source of their French toast and scrambled eggs. It felt good knowing that some of my food was coming from such a healthy, happy source. The flock was less work than a house cat and cost less to acquire, set up, and feed than buying a new iPod. I didn't understand why every backyard in America didn't have a flock of its own. They were quieter than any of the neighbors' dogs and just puttered around the garden eating slugs and bathing in clouds of dust. I envied them every day I drove off to work.

At dusk, when I returned home, I spent more quality time with the birds. Right before dark is when they're the most active and fun to watch, so I'd go out with my fiddle and play to the crowd. I wasn't very good at first, but they never complained during those early squeaks and squawks. Annie and Jazz watched from the kitchen window, tails wagging. Some nights in July, the farm was an absolute paradise. The cool Idaho summer night had me wrapped in a warm fleece jacket while hens hopped around the backyard. Mountain music wafted from my beginner fiddle as the tree frogs and crickets started their backup tracks. The honeybees hummed as they headed home to the hive from the garden, which was rich with fresh vegetables and bright sunflowers. The sun set behind the Selkirk Mountains in a pink-and-purple western sky. On those nights, it felt like everyone and everything was in its proper order, living together in my own peaceable kingdom.

HATCHING A PLAN

If you're seriously considering getting a few layers of your own, find out what type of hens will work best for you. Take all the variables I've mentioned – climate, space, and eggshell colors – into account when making your decision. Many bookstores and feed stores offer breed reference help, and if you're not too shy, ask the people who are buying chicken feed what types of birds they have. (There's a great chance that they already did this kind of figuring for you.)

There are three main ways to go about acquiring chickens: adopting or purchasing birds from a local farmer, ordering day-old chicks from a mail-order hatchery, and ordering young started pullets.

HENS, READY TO LAY

Unless you happen to have a chicken-farming friend or coworker, you'll have to do some digging around to find quality birds. The two best places to ask are farmers' markets and gourmet restaurants. This might seem like an odd coupling, but hear me out. The reasoning is simple: farmers' markets have farmers, and many gourmet restaurants are joining the locavore craze, which means they need a local source for their eggs. Try the markets first. Find people selling free-range eggs and start talking shop. Ask what kind of hens they have and why they chose them. More than any other kind of hobbyist I've met, poultry folks love getting new people involved. As soon as you give them the green light to talk birds, they'll be telling you about coop design plans and winter feeding regimens. See if they'd be willing to sell you a trio of hens. If you have no luck at the market, call some high-class eateries and see who their egg suppliers are.

Adult birds need very little time to get used to their new digs, but there are still some tips that make it easier. The most important thing to do is to have their coop ready for them when they arrive. Wait until it's dark and set them up on their new roost. Chickens really do take better to their new homes if they wake up in them the following morning. If you live somewhere rural or have an open backyard, don't be afraid to let them live without a pen. The birds always return to their coop at night; all you need is a sturdy latch on the door. When they've all filed in, close it and your work is done.

PEEPS IN THE 'HOOD

Ordering chicks means you'll know the birds from day two or three and be able to watch them grow into feathery adulthood. You'll be able to pick exactly the kind of chickens you want, and in just a few months they'll be laying eggs and napping in the sun. The downside to this is that you do have to wait several months for eggs, and for those of us excited to have a healthy, fresh source of local food, raising chicks might be a sentimental extravagance. It's a really cute sentimental extravagance, though. I still have two of the Silkies I raised from chicks, and they're sitting on a pile of eggs right now. I might have a whole new farm-raised bunch of little babes soon, and seeing them grow to parenthood is nothing short of touching.

If you're thinking of starting with chicks, request a bunch of catalogs from hatcheries. Part of the fun of planning a flock is paging through those colorful catalogs and picking out the birds that suit you best.

Ordering pullets means you get young healthy birds, and your first eggs are just a few weeks away. The compromise is you don't get that bonding and nurturing time of raising them by hand, and you have to choose from a much smaller selection of birds. Usually the only breeds sold as started pullets are Rhode Island Red crosses called Production Reds or Red Stars. Occasionally a hatchery will offer White Leghorns or black production birds, but that's pretty much it. They come on specific mailing days, which you'll need to sign up for, and you'll have to pick up your animals in cardboard carrying cases at the post office. I was a little taken aback when I picked up my chicken and saw her beak had been clipped. In large hatcheries, the points of pullet beaks are cut off to keep them from hurting one another in stressful times. Being used to farm-raised chicks from Diana's place, this threw me off. But despite this one cosmetic (and I imagine painful) flaw, Mindy's the best layer I have.

FEED, WATER, REPEAT

Adult birds eat the same basic feed as chicks do, but with more calcium and other adult-bird proteins and nutrients. I have four grown laying hens and one royal rooster, and a fifty-pound bag of feed lasts me a month or longer. If you're going to all the wonderful effort to raise a flock at home, I strongly suggest you spend the extra five bucks a month to purchase organic feed. A chicken's egg isn't organic just because it was laid in the backyard. What you feed your birds matters, and if the grains in their generic feed were grown with pesticides, your birds take that in. Support the organic feed market; you'll feel better about what's frying in the pan, and the birds will be healthier for it.

In addition to feeding your flock regularly, keep a dish of crushed oyster shells available at all times for them to pick at. It gives the eggs strong shells that can take a drop on the grass or a roll on a countertop. Hopefully, the chickens will have a chance to roam and explore a yard or garden. This lets them add special proteins and plants to their diets. Chickens are true omnivores and will eat anything they can get their beaks on. I've seen my birds take on frogs, voles, and mice, as well as grasshoppers and ants. They also follow me around the yard, eating grass clippings from the push mower as I putter on by.

FROM EGLUS TO DOGHOUSES

Your chickens will need a warm, dry place to call home. If you don't own the home you're living in, you're probably not expecting to stay there as long as the life of your birds (and chickens can live up to ten years!), so coops that can be picked up by one person or moved on a dolly are ideal. When you've got such a small number of animals, your options really open up. You can build your own garden coop, buy one from a supplier, or use a little resourcefulness. For example, a standard doghouse with a removable roof is perfect for a trio of laying hens. If you think that's too small for three birds, remember — a coop is not a cage. It's simply where they roost at night and where they lay their eggs. All you need to do is raise it off the ground. Keeping the birds a few feet in the air protects them from ground predators, flooding, mud, and snow. It's a simple insurance policy that dates back as far as most chicken farmers can remember. Use a few sturdy cinder blocks, and add a little ramp so the birds have an easy time getting in and out. Line the floor with cedar-chip bedding and straw and secure a door over the entrance to keep out the wind and any rowdy raccoons and dogs. I suggest a simple latch to ensure that predators can't break in. There you have it — a home for your birds.

Alternatively, you could buy yourself a backyard chicken coop. Since the urban-flock trend has been picking up, there are a few companies that make garden coops and runs with just a few birds in mind. The ultra-hip Eglu is one example. While it's a bit on the pricey side (around five hundred dollars), it really is everything a few hens could ask for. I ended up keeping my birds in something that falls between the makeshift doghouse and the Eglu. It's a basic coop called the Chick-N-Hutch, which is delivered in a flat box, assembled with screws, and can be disassembled in less than twenty minutes and put back in its box for moving.

I have my pair of Silkies raising a family in a small coop made of an old metal drum, apple crates, and some metal siding. It's not pretty, but it's warm in the winter and holds still during fifty-mile-an-hour winds. After the worst rainstorm, the inside hay is dry enough to use as campfire kindling. Whichever kind of coop you decide to build, make sure the birds are dry, protected from the elements, and warm or cool enough, depending on your climate.

Whether you buy a prefabricated kit or construct your own house, keep a few things in mind. Remember that predators can be a problem, even in cities. Rats, raccoons, or curious cats and dogs can wipe out your birds if they aren't protected. The coop should have a latched door you open manually every morning and close at night, or your birds should be enclosed inside a safe pen. If you have a fenced-in backyard or live in a rural area, a pen may not be necessary. In a town or more densely settled area, a penned run is a must, to keep the birds from wandering into traffic, not to mention your neighbor's yard.

Chickens need room to walk around outdoors in the fresh air. Just like hutches, prefab foldable pens for backyard flocks are available, but a few pounded T posts and some chicken wire will do the same job for a fraction of the cost. My birds used to have free range in my yard and fields, but neighbors complained about the girls strutting onto their lawn, so I had to restrict their yard time to evenings, when they wouldn't wander as far (they tend to stay pretty close to their roosts in the late afternoon). I made a large run and placed some flight netting around the edges to keep the birds confined while I was at work. When I come home from the office, even before I go inside to greet the dogs, I open the pen's gate and let the birds file out to scratch, crow, and explore.

CARING FOR CHICKS

On their first day home from the post office, chicks need immediate attention. First of all, you should have a brood box ready for them when they arrive. This is a container kept at a constant temperate of 90°F. A sturdy cardboard box with a lightbulb dangling from a cord is as fancy as you need to get. Line the box with some cedar shavings, and have a small font (a watering device) and feeder ready and loaded. When those chicks come in, they'll be parched and starved. Gently lower them, one at a time, into the brooder and make sure all new arrivals stop at the bar for a nosh and a sip.

Food and grit. Chickens that are not yet laying need to eat a basic diet of chick ration. Bought at feed stores or online, it is specially formulated for growing birds. Think puppy chow for the feathered set. Basic bird kibble isn't enough for small chicks, though. They need some roughage with their feed to aid their digestion. This is where chick grit comes in. Since chickens do not have teeth or an exceptionally strong digestive system, they need some pebbles in their diet, to help grind up food in their gizzard. I suggest adding corn scratch and grit that's specially formulated for chicks.

Warmth. The biggest reason for chick death is cold. A frigid night or a strong draft can kill one just as quickly as a curious dog can. Without a mother hen to nestle under, they need you to give them a warm home. Monitor them as often as you can. You'll know the birds are okay if they're wandering around the brooder box freely, unbothered. If they're huddled under the light, you know it's too cold for them. Lower the bulb a bit. If they're pressed against the edges of the box, they're too hot and the bulb needs to be raised. You'll get the hang of it in no time. After the first week, the bulb should be raised to decrease the temperature by five degrees each week.

The chicks should stay in a controlled indoor environment like the brooder until they start "feathering out." In three to six weeks, they'll go from adorable poofballs to these hideous, gangly, awkward proto-chickens. Stand by them — you had an awkward period, too. If you have an insulated coop with a roost and a shaving- or hay-lined floor, feel free to move them outside, as long as it's not the dead of winter. (You might want to keep them in the garage if there's still snow on the ground and your coop is on the small side.)

CHOOSING YOUR BREED

When choosing your breed of laying hen, there are a few things to consider. The main variable is your climate: if you live in Maine and know cold winters are inevitable, consider a heavy, hardy breed like a Brahma or an Orpington. If you live in the Tennessee Valley, you might want to opt for Leghorns or less-hardy birds with fancy combs. You should also think about the color, size, and quantity of eggs you'd prefer. If you only have a small space to keep chickens, you might choose small birds. Chickens are flock animals, so if you can make four Bantams happy in the space two full-size birds usually get, it might be a better choice for all of you. If you want some color in your collecting basket, consider getting an Ameraucana. These birds lay Easter eggs in blues and greens. A Welsummer lays an egg so dark you'd think it was made of milk chocolate. If you're being practical and just want lots of fresh brown eggs, your best bets are the classic Rhode Island Reds and the Red Stars. These are hardy birds and great layers.

URBAN BIRDS

Chickens are wonderful. They are clean, cheap, and quiet. They take up less time than a gerbil does, and the eggs they lay in your backyard make the ones in the grocery store taste like garbage. If I haven't sold you yet, they're also becoming trendy in cities and suburban backyards around America. From Brooklyn to Boise, city flocks are popping up in places where sidewalks and streetlights can be seen from the coop. Most cities, probably yours as well, allow hens in open-air spaces if they have a concealed coop and run and you don't have any roosters to wake the neighbors. Chicago, New York, Seattle, Portland, and Memphis, to name just a few, all have urban chickens. There are Web sites and suppliers dedicated to the city chicken set, offering orders of as few as three chicks and futuristic backyard coops that are made of plastic and can be quickly hosed down for cleaning. By checking with your county clerk or doing a search online for local legislation, you'll quickly find out if you, too, can have a few birds to play music for.

GROW YOUR OWN MEAL

I get cocky with my hoes

MY FIRST WINTER in Idaho was pretty typical. It snowed. But I'm not talking about country-cottage snow globes here, I'm talking about snowdrifts so high I couldn't see out the garage windows. Having grown up in Pennsylvania, I was moderately acquainted with the stuff. But when snow came there, it was quick and violent, often in winter thunderstorms that passed in a night, rarely accumulating over a couple of inches and then fading away. In Sandpoint, snow came like a trotting Percheron: beautiful, steady, and strong. As it continued to fall for weeks without melting, it piled and drifted and froze all around the farm. I felt like I was meeting winter for the first time.

Being snowed in gave me a chance to plan summer projects, which at that point I romanticized and overestimated. I imagined summer on the farm to be Eden. I pictured smiling dogs gnawing on beefy carrots in the shade of the old birch tree and chickens running around doing interpretative dance. I'd hum in a hammock to the radio while butterflies punctuated the blue skies like the happy little clichés they are. It was so removed from conceivable reality, but that didn't stop me from planning. Out of all the isolation-induced blueprints, the most exciting was the garden.

Oh, the garden. It became an obsession. This was going to be the heart and soul of my rented freehold. It was going to be something else, all right. It would be a couple hundred feet of cold frames and tarps. Rows of hand-sown seedlings would bloom and pour over the railroad ties, a bounty nothing short of paradise. Neighbors and passersby would slam on the brakes in awe of my creation.

However, there's a huge difference between reading about a vegetable garden and actually planting one. I learned this the hard way. I had never started my own homegrown vegetable operation before, and I was getting drunk on the idea of it without realizing how much effort it would initially entail. I was planning what I then thought would be a manageable operation of two hundred square feet overrun with sweet corn, tomatoes, broccoli, squash, zucchini, potatoes, lettuces, and peas. (Now the idea of one person with that much zucchini kind of nauseates me.) It was going to be my own glorious produce section right there in the backyard. I prepared for weeks. I collected seed catalogs and checked out all the gardening books the library permitted. Within a few weeks I had rounded up enough books, potting soil, and hand tools to open a gardening kiosk in my kitchen and supply half the county with reference materials or, at the very least, incredibly informative bathroom reading.

Finally, after months of snow, thaw, and mud, the soil by the barn was ready for the wrath of my hoe. I pounded some stakes in the ground and roped off my two hundred feet to freedom. It looked like nothing, no sweat. Not even as big as a standard swimming pool. I steadied my footing, raised my hoe in the air, and started hacking away.

Let me tell you something. Hoeing is really hard.

After about two hours in the April sun, which wasn't even hot to begin with, I was panting like an ex-racing greyhound trying to sprint around the track after four months on a futon. My carelessly ungloved hands were blistered and splintered, my back ached, and all I had prepared was a small rectangle. It was about five feet by three feet. I had barely made any progress. I was about ready to throw up. Let's hear it for me.

I hadn't realized how much effort goes into breaking sod. Just to clarify here, sod means grass and grass means roots. To get the desired result — an eighteen-inch layer of loose rock and root-free soil — I had to remove this dirt-acne called grass. That flew in the face of everything I had understood as someone who grew up in a town with backyards edging against other backyards. Lawns were something to be prized and to cause envy in neighbors with lesser aspirations. I had been taught my whole life to maintain and cherish grass, and now I wanted it off the face of the earth. In the fray of the hoeing, I saw lawn as a silly toupee that ruins perfectly good food-growing soil. The grass below my feet hadn't been touched in two decades, save for mowing. That means it had spent twenty bitter years surviving in a place so cold that I had to scrape ice off the *inside* of my car window in January. It wasn't about to give up on its home just because on that particular morning some girl from another state felt like it ought to.

Sod breaking went like this. First I had to pierce the sod with a shovel and then pick it out with the hoe. This required a lot of muscle and several attempts of beating it into submission before it gave in. When I finally broke through, I had to keep hacking away at the topsoil till I hit clay, rocks, roots, and bugs be damned. When I wasn't hoeing, I was chucking stones and yanking roots. I'm far from a delicate lady, so I was fine for the first twenty minutes. But then I started to ache. Little pains started to creep into my arms. My shoulders started to gossip with my back, and half an hour later they both resented me. I kept trying with all my might to dislodge the roots that had shot back to life every spring since *The Wonder Years* first aired, but they were tough customers and I was a girl who planted window boxes. I called the sod some pretty horrible things.

After two hours of this, I couldn't imagine being able to take any more that day without dislocating something or ripping my hands open. So I stopped, gave the rest of the area I had plotted a good long look up and down, and promptly gave up. I had been defeated in honorable combat. That grass hated getting stabbed and fought back with everything it had. Forget about Texas; don't mess with Idaho.

And that, my friends, is how I ended up with three small raised-bed gardens. I scaled back down to reality and I'm glad I did. The three plots were easier to manage, cheaper to fence and cover, and let me plant things in places that made sense. In less than fifty square feet, I was able to grow all the types of food I had originally planned on, with enough to give away to friends. Maybe someday I'll have that Eden I'd dreamed of, but for a beginner with angry sod and hand tools, it was just too much. Just between you and me, I still think my original plans were solid (if only I had been able to find a team of Amish kids and a rototiller).

I DIDN'T THINK about all this back in January when I was perusing seed catalogs in my warm kitchen. Sure, I figured there'd be a lot of initial work, and some weeding and watering once the plants were in, but I stubbornly needed to be a source of some of my own food. This might be the single most satisfying accomplishment the human animal can achieve. Food is one of the few things we all have in common, the thing that sustains bankers and beggars alike. If you can garden, you're literally giving life to what sustains your own.

Besides, in the quest to eat local foods, you can't get more "local" than twenty yards from your kitchen sink. And all the ripped-up hands from sharp weeds and sore shoulders from hauling water make that tomato sauce and simple salad on the deck at sunset so much more fulfilling than what's on the grocery store shelves. You just can't know till you're forking it in for yourself.

Deciding what to grow was challenge number one. Let me tell you, being practical wasn't easy at first. I started off treating the garden like an edible installation piece instead of a reliable food source. I wanted my own whimsical placement of colors, shapes, and sizes, but didn't always think about what kinds of conditions the plants needed. If I planted giant sunflowers in the center because they look pretty, they would cover the peppers in shade for half of the day — not good, since peppers need crazy amounts of sunlight to produce. If I planted my favorite herbs, mint and lemon balm, right in the garden instead of in their own containers on the porch, they'd become invasive and choke everything in their way.

Eventually, I learned to plant the tall stuff on the northern end of the garden so it would never shade what was below it. I learned to stick vines and crawling plants at the ends of the rows, so their gourds and melons would grow outside the beds and not overwhelm the plants on the inside.

I also came to realize that when you're starting your first garden, it's probably better to grow a lot of four vegetables you know you'll eat than to grow twenty vegetables that look beautiful but you don't eat all that much of or have never tried. Yeah, I made that mistake. Those seed catalogs are chock-full of gorgeous, colorful things, and as a designer all I wanted to do at first was make the garden as beautiful as possible. I was so in love with the idea of having this big, sprawling vegetable patch, I planted all these squashes and zucchini. They grew fine, and sure did look pretty curling around the rows. The problem was that I hardly ever eat squash or zucchini. When I'd eaten all the broccoli, I started to wish I had managed my real estate a little better. I had bushels of a vegetable I wasn't psyched about and not enough of the ones I really wanted for dinner.

Even still, during that first summer I was able to have a meal a day right from the backyard. My three small raised beds grew jammed full and kept me fed for almost two months! The fresh veggies made salads and stir-fries, and the laying hens' eggs and homemade jams always complemented my breakfast. The excess I froze or canned. When out-of-town friends passed through on their way to Seattle, they got to eat whatever was ready for harvest. If you think making yourself a meal from your backyard sounds satisfying, try feeding friends a meal you grew *from seed.* That wins, hands down.

. . .

VEGETABLE GARDENING has been called "the peaceful sedition" because at the most basic level, when a person can feed and shelter herself, she doesn't require a government to provide for her. (Don't get all fidgety, now — this isn't about raising anarchy flags and minting your own currency. But hey, if you do start making your own money, can you put my face on the twenty-six-dollar bill? I think that would be awesome, thanks.)

It's not just about pride or independence, or even connecting with nature. It's about wanting hash browns on a Saturday morning and being able to run out to the backyard in your bathrobe to grab some potatoes from the garden. With a minimal amount of effort on one Saturday afternoon, you can set up a garden (either in the ground or in containers) that will feed you for months to come. Growing vegetables is the most practical skill you can pick up, and definitely one of the most rewarding.

After my first beds of veggies really started cranking, I'd go outside, put my hands on my hips, and survey the gardens with a giant smile on my face. I'd look at the rows of potatoes and onions and run menus through my head. When the rewards of your work are heavy on the vine and you're deciding between stir-fried broccoli and grilled sweet corn for dinner, you feel the satisfaction of knowing that everything you need is sprawling out before you and thriving. And you're both responsible for its existence and the one who benefits from it. The world begins and ends here. I don't know what else can fill up a person's soul as much as that. I guess starting a major world religion might feel the same, but I don't think on such a big scale. Snap peas will do. Besides, peas involve a lot less paperwork.

YOU CAN DIG IT

When the desire for homegrown food overwhelms your preference for clean fingernails, there are a lot of ways to get started. You can fill a large container planter with tomato plants from a local nursery or plant organic seeds in tilled soil. Whatever your aspirations, know that the following bits are general advice and over the next few months you'll want to do more research of your own on what makes the perfect garden for you.

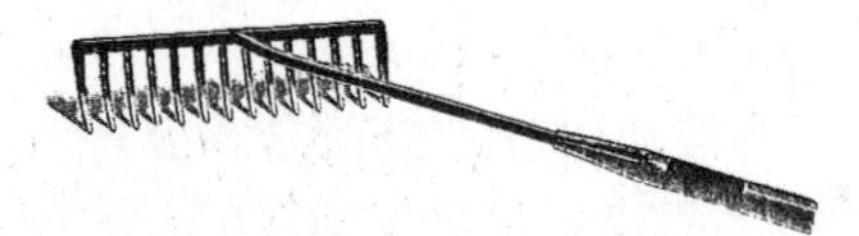

YOU ALWAYS WANTED A SIX-PACK . . .

When you're first getting going, you can choose to start from seed or you can purchase started plants (these are seedlings, usually sold in packs of four or six at your local nursery or hardware store). Although buying six-packs means you'll get less of a selection than if you started from seed, it's much easier, especially if you'll only be growing a few plants. Also, there is a much lower chance of failure than if you start from seed. I suggest trying both if it's possible. You just may find yourself in the depths of February yearning for something green and growing, and starting seeds is just the trick.

When you get ready to transplant your seedlings in the spring, you'll want to harden them off, or gradually acclimate them to life outdoors. Start by taking them outside to a shady spot for just a few hours and then bring them back indoors. Increase the amount of time and exposure a little bit each day, working up toward putting them in full sun for several hours, then leaving them out overnight on a sheltered porch. Let your plants have a week or so to get used to the change before you dig into your containers or tilled soil.

MAKE A PLAN, SON

Even the simplest garden requires a game plan, and your garden of containers or beds is no different. The first step is to decide what it is you like to eat, and eat fairly often. The second step is to check around and find out what varieties are best suited for gardening in your area. For containers and small beds, there are special versions of your favorite veggies that are designed to grow the largest and heaviest fruits on the smallest plants. Consider a 'Celebrity' bush tomato for your planter. A two-foot-tall plant can produce tomatoes up to twelve ounces apiece! There are also all types of size-specific hybrids. People have created round eight-ball-size zucchinis and watermelons as small as eggplants.

Spend some time online and order several seed catalogs to peruse all the different varieties of edibles. You'll be able to find everything you need to get started, but please be warned: Catalogs are enablers in what will surely become a vegetable-growing addiction. When you start seeing hydro countertop systems and indoor heated seed-starter greenhouses, you'll end up deciding you can take on more than you can handle. Remember, just 'cause your thumbs are green doesn't mean your horns aren't.

Once you have a few hearty veggies chosen and seeds ordered, it's time to start prepping indoors.

A container garden does have advantages over an in-ground plot. You don't have to worry as much about animals and insects, and plant diseases are considerably fewer. But you're not home free. You can over- or underwater your vegetables. You can set them in a spot with too much sun or too little. You can put them in soil they can't use because it lacks the special needs of a particular plant, while the bag of nitrogen-rich soil they need sits on the store shelf. The best thing you can do when you're getting ready is ask for help. If you're willing, take a trip to a nearby greenhouse or nursery and ask questions about your space. An expert on the local climate can be a priceless resource. Another choice is to hit the books (see Research, Son, page 178). Grab a small-space gardening book or one that's specific to the kinds of veggies you're drawn to. While you don't have to have in-depth intelligence to get started, little tips and facts you come across along the way can be invaluable.

It's really important that you buy potting mix for your containers — don't just scoop dirt from the backyard. Bagged potting mix has a combination of water-retaining peat and water-repelling perlite for good drainage (regular garden soil will compact in a container). Mix together two-thirds potting soil and one-third organic compost and put into large planters that have drainage holes. Throughout the growing season, you can top-dress the containers with more compost, to provide extra nutrients.

Digging in (or not). Two words: *raised beds.* Raised beds are a saving grace for new gardeners. They're small, manageable, simple to construct, and provide good drainage, which plants need to thrive. What exactly *is* a raised bed, you ask? It's just what it sounds like: a bed of improved soil that's mounded above ground level, generally between eight and twelve inches high. Some people like to build a wooden frame for their beds to keep them all neat and tidy, and you can, too, but it's not strictly necessary.

Choose a spot for your beds that gets at least eight hours of sunlight a day, and plan on siting the tallest plants (corn, tomatoes, climbing vines) on the northern side and sprawling vines (pumpkins, squash, cucumbers) on the southern side, in a spot where they can creep over the sides of the bed and away from it. Plant north to south because the sun will cast a shadow over everything in the taller plants' way if you don't.

As a beginner, it's a good idea to start small and successful, maybe with just a single bed, two feet by four feet. If you're feeling frisky, aim for two, but for new gardeners without a chicken army to bail them out in times of crisis (see There's Not Enough Salt in the World, page 45), this will be plenty to keep you busy. Here are a few tips to get you going:

Rather than spending a lot of time digging up sod, mow the grass down to the nub and cover the space with several layers of newspaper or brown kraft paper.

Treat your raised bed like a giant container and fill it with a mixture of half potting mix and half high-quality compost. (Don't buy the cheapest bag you see. You get what you pay for.)

If you like, line the edges of your little garden with wooden planks, scrap wood, bricks . . . anything, really. (Well, anything nontoxic. Keep away from materials like pressure-treated wood, which leaches chemicals into the soil.) You can even nail together a simple box frame to keep everything tidy.

Plant seedlings that have been hardened off (see page 38). Water them with a gentle, but thorough, sprinkling. A watering can with a fine rose (sprinkler head) works well. You want the earth to stay moist but never be flooded.

Give plants some extra protection from sun, wind, and cold those first few days they're in the ground. Cover them with old baskets or a thin sheet (something that will allow light to filter in) for a day or two, until they're used to their new digs.

Make sure your outdoor space is prepared for you to enjoy it! Put up a hammock or bring out some blankets and a book. You should be able to relax after all that hard work.

WEED, WATER, REPEAT

As you get more adept at gardening, you'll learn all kinds of techniques to improve your harvest. You'll pick up tips like what types of vegetables do best in your region and how compost tea can heal a world of troubles. For a novice, though, it's wiser to stand back, observe, and let nature run its course.

If you planted your little gardens to the best of your ability, with common sense and good advice in mind, the only work ahead is to keep the plants alive and growing until they're ready to eat. This means occasional weeding, watering, and providing protection from animals and insects that might harvest them before you do. Water your plants daily in hot weather, but not for at least a day or two after a heavy rain (too much water can be just as bad as too little). If you've started with quality compost, your plants will thrive. Some, like corn, require more of certain nutrients than compost offers, and they'll need a dose of organic fertilizer once a month or so. You'll learn the nuances as you go. (For more specifics, check out the books in Research, Son, pages 183–184.)

Every week, take a few minutes to pull out the weeds that grow among the vegetables. Weeds compete with veggies for nutrients and water, making them smaller and less productive than if they had been grown in a weedless environment. Keep out neighborhood children and other small varmints with a fence. I made a fine little fence out of inch-thick dowel rods and some tree netting. It looked awful, but it worked great. Big critters like deer require more reinforcement (but this is a topic worthy of an entire book — see Research, Son, pages 183–184).

The commonsense catchall to caring for your garden is simply to pay attention to it. Check on it the way you would a goldfish — with mild interest and basic care. Unlike a pet, though, if your plants die, there's really no harm done. Just pick up a few more six-packs from the farmers' market and try again. No sweat.

NEXT STEP: BEES AND CHICKENS

Here's a small farm ecosystem that works wonderfully: vegetables, bees, and chickens. Even the smallest backyard has space for a hive (see Beekeeping, page 49) and a three-hen coop the size of an average doghouse. Honeybees and hens are masters at helping a garden thrive, and in the process of doing so, they supply you with all the eggs, honey, and entertainment you'll need all year.

The bees are expert pollinators. It's what they do, and they do it well. Having even a small hive means your gardens, fruit trees, and flowers will be bountiful. And hens are four-star pest removers. They patrol your rows of veggies, scouting for slugs and bugs, and happily chomp them up while ignoring the bees. As long as the chickens are out in the garden when you are, you can make sure they don't start snacking on your produce; but generally, being the hardy omnivores they are, no chicken is going to settle for your romaine leaves when it could be dining on slugs. Personally, I would opt for the romaine, but I seldom try to understand chickens.

FROM POOP TO PLANTS

One advantage of keeping chickens (and rabbits!) is that you can use their waste for your garden. When you clean out the chicken coop or rabbit cage, dump the old bedding and poop into a pile in a (perhaps distant) corner of the yard and let it rot away. Throw in some kitchen scraps and yard waste, let it rot, and you've got homemade compost. It's a good idea to keep an eye on the ratios of what you're adding — ideally, it should be about eight parts brown stuff (straw, wood chips, dry leaves) to one part "green" stuff (grass clippings, rotting vegetables, chicken and rabbit poop) — and to turn the pile with a garden fork now and then to get some air mixed in (gotta keep those microbes happy). It can take anywhere from six weeks to a year for waste to break down, depending on what you add to the pile and how often you turn it. When it's brown and crumbly and smells like earth, not garbage, you're golden.

. . .

Slugs. The little slimy terrorists were eating through my romaine as if they'd signed up for a farm share program. I tried a few tricks people had suggested. I put pie tins of cheap beer between the rows for them to crawl into and drown, but it didn't work as well as everyone said it would. Maybe Idaho slugs have some hesitation when it comes to certain libations. Or maybe the jerks in my garden have a slimy aunt in AGA (Alcoholic Gastropods Anonymous) and are hesitant to drink up?

After the beer bit, I sprayed with organic pesticides. That worked okay, but I had to reapply the mixture twice a week, and with all the plants I was growing, a twelve-dollar bottle barely made it through two coatings. It seemed like a lateral move to spend as much money growing food as it would cost to buy it, so I went with a more hands-on approach. Before work, I went through every plant, leaf by leaf, picking off the slugs. This was by far the most effective method, but to really do it right took thirty minutes for each raised bed. I didn't have two hours a morning to play with slugs. I had dogs to feed and coffee to drink, both much more enjoyable than getting my fingers slimed at dawn.

I was at my wits' end. I didn't want to lose my gardens to such a small adversary, but I couldn't afford to keep spraying, and handpicking them was becoming so stressful I was considering walking through the produce aisle of the supermarket just to relax. I was starting to doubt the pictures on my seed packets. Nowhere on their labeling was there a yellow sticker warning "Jenna, you're going to have a small, hungry army take over your scrappy wannabe-gardener plots, so you should probably quit your day job and dedicate all your free time to these salad greens if you want to see the goods pictured here." (It's a large sticker.) I was getting desperate. I needed another slug-control option.

In the end, I found one. It happened on an exhausting Sunday afternoon after two long hours of hunchbacked de-slugging. I stood in the shadow of the old barn and surveyed my little homestead — my garden beds, the old fences still wet from the morning rain, the chickens puttering around the walkways, the house. My head snapped back to the Black Silkie Bantams taking turns trying to de-corpse the yard of worms. Chickens? Hmmm . . .

I had heard that chickens make for great pest control. I picked up my five young, cooing Silkie Bantams, carried them in both arms across the farm, and put them inside the fenced gardens to have at it. I brought out a book and a blanket and read while they hunted for the slugs. After twenty minutes in a bed, I carried them over to the next one, until an hour (and three chapters of a novel) later I was well read and they were well fed. I did this twice every weekend and the holes in the leaves stopped appearing. I didn't have a serious slug problem the rest of the summer, and I read two books while tanning out in the garden. Now that's teamwork.

ON BORROWED LAND

The issue with most landlords isn't the idea of your growing food. After all, who wouldn't love a nice, quiet gardener for a tenant? What turns them off to the hoe-raising idea is what a crazy, fresh-tomato-lusting renter is going to do to their landscape. If you're lucky enough to have outdoor space that the sun hits, you've got a chance to create a small (but very productive) produce mart. If you're in an apartment and have no chance of planting in the ground, you still have options. You can use planters, window boxes, or even install little greenhouses to hang outside the windows. (These are simply glass extension boxes you install in a normal window to make a little bubble of a greenhouse.) You could ask to use the roof. (Hey, if you're planting up there, maybe you can sweet-talk your super into allowing a beehive or two.) Even the most modest apartment with a sunny windowsill can host at least a container of herbs for tea or cooking.

For some people who are just itching to get down in the dirt, container gardens won't do, and that's a problem if you're living in a studio apartment on the fifteenth floor of a high-rise. But there are still options. There's a good chance you know someone with an outdoor plot within walking distance of your apartment: small businesses, friends, or neighbors, with yards, courtyards, or rooftops. Any place that can take a load of compost can start producing food.

Even if you don't know your neighbors, it never ever hurts to ask people if you can garden the ground they aren't using. Honestly, the worst they could do is say no or look at you funny. (And since when has that ever stopped you?) So rally up the moxie you need and ask if you can plant a beautiful garden in their ground. Tell them it will not only beautify their space, but you'll be willing to give them 25 percent of the yield as payment for use of the land. You're not likely to be turned down then. It's a win–win for everyone.

BEEKEEPING

Watch the new kid break out in hives

LIKE MANY OTHER WOMEN, I've spent a fair amount of time dreaming of the day when I'll be dressed in flowing white and a veil. I'll be outdoors on a beautiful sunny day, surrounded by flowers, open fields, and the buzz of eager activity.

But this wasn't exactly what I had in mind. I was behind the farmhouse in early May, in a ridiculous, oversized white beekeeping suit, complete with mask, gloves, and a smoker, which is kind of a miniature tin woodstove that looks like an accordion on Demerol. I was hunched nervously over the beehive I had assembled in my living room the night before.

Here's a hint for you fine readers thinking about beekeeping: They don't send you the hive assembled. You get a big box with seventeen jillion little wooden slats, combs, and nails. While it wasn't hard to figure out how to put it together, opening that box was a little intimidating for a girl whose toolbox came from Ikea.

When everything was in the shape of a hive, it looked like a filing cabinet for bees: a rectangular box with file folders of combs that hung from little rafters. It smelled like beeswax. I kinda dug that part.

Besides bee home construction, there was other business to be prepared before installing the bees. I had to stock their cupboards with food. So I bought a box of pollen substitute, a yellow powder you could make into little patties and place in the hive or sprinkle around outdoor plants. I also had a pot of water on the stove with half a pound of sugar simmering in it to create a syrup starter food. It would be placed in an upside-down Mason jar with a perforated lid right at the hive entrance, where worker bees could help themselves to room service if they didn't feel like going for takeout. (New colonies appreciate all the help they can get.)

Anyway, bee foodstuffs and carpentry accomplishments aside, I was about to welcome my new tenants into their hive any minute now, and I wasn't playing it cool. In my shaky hands I clutched a roaring box of Italian honeybees. Now, I usually pride myself on not being bothered by bugs. But there's a huge difference between the minor threat posed by a wasp landing on your picnic platter and the very real possibility of getting so many stings that you could slip into a coma. At that moment, no one in the world could convince me there wasn't a fine line between hobby farm enthusiasm and a death wish. Too late to turn back now, though.

YOU MIGHT BE WONDERING, why all the looming trepidation if I was wearing the whole beekeeper getup? Before I got into this mess, I always assumed the beekeeper uniform was a sting-proof barrier, a surefire shield that stood between me and the Epi Pen. Guess what — it's not. Some people think the color white lulls them into complacency. Mostly, it's meant to keep bees from crawling in under your clothes. Truth is, the thin cotton suit couldn't protect me from being stung even if it wanted to. Great.

Despite the risk of stings, beekeeping was one of the main things I wanted to learn when I first moved to my Idaho homestead. Honeybees are so much more than insects that give us the sweet stuff. They are a complex society with a class system, daycare facilities, entourages, and funeral processions. The more I learned about them, the more enamored I became. Apparently, they are also a gardener's best friend. Being new at gardening, too, I figured this summer I could use all the help I could get.

SO, THERE I STOOD, the moment of truth just minutes away, with 10,000 angry women who had no idea what they were doing in Idaho. (At that particular moment, I can't say I did, either. There was some bispecies camaraderie in the chaos.) I had no idea what was going to happen when I took off the lid and dumped them into their new home. The how-to books were all sort of vague at this part, which I thought was a bit shady. I mean, this is the stuff people holding a roaring box of bees *need to know.* Would the bees fall into the open hive box with a calm, humming thud or would they bitterly fly out in every direction, trying to take out their captor in a kamikaze death brigade? All I knew was that I had to get the packaged bees and their queen into their new place and close the lid.

I cursed some otherwise decent authors while I blew little puffs of smoke at the box of buzzing hordes. Using my hive tool, I started prying open the lid of the wooden box that contained the bees. I sucked in all the air around me, and on the count of three, I knocked the box against the ground to dislodge the bees from their wire roosts, then dumped them into the hive. You want to know what faith is? Faith is beekeeping without a clue.

I didn't get a single sting. Really.

The bees seemed too preoccupied with building their empire to care what the crazy lady in the baggy outfit was doing with her free time. They entered the plastic-and-wooden combs with a muted energy, like animate puzzle pieces dutifully fitting themselves into place. In another part of the hive, a huge mass of bees was lumped together, squirming around a central point. Local beekeepers told me that this is common — it means the queen is inside the huddle, wearing this suit of live body armor.

I wondered if *my* suit was calming them. If not, at least it was starting to calm me, even though I knew it wouldn't save my skin if the bees went AWOL. The suit did seem to have an effect on them. A stray bee or two would fly out and land on my arm, I'd gently flick them away, and back they'd go into their combs.

This wasn't so bad. It was actually kind of fun, standing there watching the bees mill about. It was pretty damn satisfying knowing that I had come this far. Every beekeeper I had spoken with over the past few weeks assured me that setting up the hive would be the hardest part. Feeling like the coast was clear and that Ma Nature would take over, I relaxed. I could just picture it: On weekend afternoons, I'd sprawl on the grass and watch them fly over me. With my arms behind my head, staring up at the sky, I'd see them fly over me, their legs yellow and heavy with the good stuff. In a few seasons, I'd have jars of gold to spread on toast, mail to friends, or sell at the farmers' market. I closed my eyes in the Idaho sun and took a nap. Beekeeping. No sweat.

OVER THE NEXT few weeks the bees took minimum upkeep and gave maximum satisfaction. They were a constant presence without being overbearing. I'd go out to my garden to weed and see a few worker bees hovering around the snap-pea blossoms. It wasn't hard to tell them apart from all the bumblebees buzzing about. Bumblebees are fat and hairy and round, like insect woolly mammoths, barely able to haul themselves from flower to flower. But the Italians are nimble and slim, with hairy manes of fluff around the chest and back. Every few days I'd walk the twenty yards or so to their hive behind the house and make sure they had enough sugar water and pollen substitute. That was the extent of my beekeeping responsibilities — just checking on things.

The job of a new beekeeper during the first year is simply to keep the bees alive and strong enough to make it through that crucial first winter. (If you're getting into bees and happen to live in a mild climate, you already have a huge advantage over us northern schmucks.) If you can get them through it, you can expect to harvest some gold for yourself at the end of the following summer. It might seem like a long time to wait, but hey, that year is coming anyway, so you might as well have some honey at the end of it.

. . .

THE WEEKS PASSED, and my life at work and with other farm projects kept me busy, so I spent less time beekeeping than I did before. I kept the sugar water stocked, and opened the lid once a month or so. It wasn't until early July that I noticed a problem. I opened the lid and saw half the bees I'd started with. I looked around the ground and inside the box for dead bees, to see if they could help me in diagnosing the problem, but they simply weren't there. For whatever reason, the bees were disappearing.

There was a lot of news that summer about "colony collapse disorder" (CCD) — bees mysteriously leaving their hives (and leaving scientists baffled). Was that what was going on? I scanned all my beekeeping books for information, and reread the magazine articles I'd collected about CCD. I logged on to online beekeeping communities, and everyone seemed as confused as I was. A few days later I went out with the smoker (by this point I had gotten so used to the bees I didn't even wear the veil anymore) and saw fewer than two hundred bees, feverishly tending the few combs they had produced. The hive was drastically depopulated; time to call for help. I phoned Diana, my mentor in all things farming, and told her what was going on. She ran through a mental flip file of all the possible bee-related diseases and their symptoms. It didn't sound like mites or mold or mice or any other problem that started with the letter *m,* so she said she'd come over and check it out. Maybe there was a simple explanation that my beginner's eyes couldn't see.

After pulling out the remaining bees desperately clutching their combs, we saw the problem. The queen herself was dead. Not just dead, but trapped in the very box she was originally shipped in. That poor hive never had a chance — they were doomed from day one, when I saw that big, queen-centered clump fall. It didn't occur to me that the workers were protecting a locked cell or, worse, trying to free the queen. But I tried not to think too much about that possibility. I felt bad enough as it was. I didn't realize that my bee supplier shipped its queens boxed, and that it was my duty as the beekeeper to release her so she could go about her business, making other bees and generally running the joint. I was laden with guilt and embarrassment. The only thing wrong with the hive was its ignorant beekeeper. I realized there was another problem that started with the letter *m.*

Me.

BUT HEY, success really does come down to who you know, right? Since my birthday was just around the corner, Diana had a special surprise in store for me. She pulled up to my front door in her farm-beaten Jeep Cherokee and in the backseat was one of her own hives, which she had planned on requeening. Requeening is the nasty business of replacing an older queen with a sprightly new one. Since there can be only one monarch in this society, the old queen has to be dethroned and the new one placed in her stead. My predicament gave Diana the opportunity to take her old queen over to my hive and slip the new one into hers. No lives were taken, and my lot got a second chance.

Along with the queen, she offered me a few started combs, complete with thousands of new bees, to revive what was left of my pathetic operation. We installed them with no problem, refilled the sugar feeder, and let the bees take over. I was back in the honey business, and finally on the right track. Within a few weeks, the new gals had rebuilt, and connected the old comb frames to the new ones. The stream across the street offered fresh water, and the fields of wildflowers were perfect for the pollen and nectar they craved. There was hope for getting them settled in time for winter. Feeling lucky and extremely grateful for the Hail Mary that Diana was able to pull off, I went back to working in the garden, checking the hive, watching the chickens, and playing my fiddle with renewed energy.

THROUGHOUT THE REST of the summer, everything seemed back on track. The hive was at its strongest, and whenever I checked on the gang, they were healthy and hard at work. For a while I actually believed we'd get through winter. We had overcome my mistakes and had a new brood growing in the combs. We even survived the influx of hundreds of the yellow jackets that were overtaking North Idaho in swarms. The yellow jackets had no problem walking right into a weak hive and making themselves at home. The defense on our end entailed stringing up bee-safe traps all around the hive. Jackets were drawn more to the carcasslike smells of the traps than they were to the soft yellow combs inside the hive. Through all these trials, through hailstorms, wind, fallen branches, and raiding raccoons, my brave girls kept on. When the end of September came rolling into Idaho, leaving a carpet of yellow aspen leaves around the hive, the worker bees were still going strong. So I felt okay leaving them for a weekend while I traveled to Tennessee for a mountain music festival. Everything was going so well, they'd be fine without me for a few days.

They weren't. When I returned, I found a tragedy. A bear had discovered the hive and, in a prehibernation gluttonfest, had ripped it apart like it was made of Popsicle sticks, scattering the hive body and combs all around the yard. There were claw marks and broken frames, and combs with giant chunks that had been bitten off. I looked all over for any sign of the bees, but they were long gone. They had swarmed off when the disaster struck, just like you or I would have done. There was no rebounding from this. With our first real snowfall just a few weeks away, no new hive would have a chance.

I wasn't angry or disappointed. I was just plain shocked. I kept going over the planning, the construction, the idea that airplanes and friends and miracles had to converge to keep a heartbeat in this little box. In one instant, another animal destroyed all of it. Not with malice, but just because of its own need to survive.

It was then that one of the many hard truths of living closer to the land hit me: All of this (me included) was at the mercy of nature and circumstance. If the coyotes had the *chutzpah* to risk an encounter with Jazz and Annie, they could come right up to the house and kill my whole flock of chickens in a single night. My dogs could cross paths with a moose and get trampled. The entire farmhouse could be flooded or slammed by a tornado. A bear could chomp on me while I was out hiking in the mountains. All these things could happen. I'm not a paranoid person, and I don't live my life expecting bad surprises to ruin everything, but I was surveying the pieces of my wrecked hive all around me. It happened. It didn't matter if that hive had been impeccably tended since day one — it was gone because some bear out there in the back wanted a snack. Like anyone else who lives in the country, backyard beekeepers have to be able to react to circumstances beyond their control and to learn from whatever life throws at them.

. . .

OKAY, SO MY FIRST attempts at beekeeping weren't successful. Other skills I learned from scratch seemed to work out for me, but I guess you can't be lucky at everything. I can't say I'm proud to admit that I spent a summer killing 20,000 bees. But hey, fail better next time, right? (I found out later that other beekeepers in my area keep barbed-wire, chain-link, or even electric fences around their hives to protect them from large animals. A note for those of you in bear country.) I plan on starting up another hive soon. One jerk of a bear isn't going to keep me from producing my own honey.

Considering what I know now, I don't think I can look at the jars of honey on supermarket shelves in the same way ever again. The planning, the margins of error, and the nakedness of a vulnerable hive . . . I've experienced these firsthand. It amazes me that we humans ever figured out beekeeping in the first place.

I will say this, though — the day I finally do harvest my first batch of honey, I will consume it with absolute mindfulness, respect, and appreciation. Drinking that first cup of sweet tea will be like going to church; there will be awe and gratitude in every sip.

STARTING A BUZZ

To get started with bees, you're going to need to invest in some basic gear. Beekeeping suppliers have complete beginner kits, which contain all the items listed below (well, save for the bees themselves, which you'll need to order from a reputable apiary). The one I started out with cost about a hundred dollars and had everything I needed to install my first colony. Not too bad, considering you can get up to fifty pounds of honey at harvesttime!

YOUR GUIDE TO BEES (ABRIDGED)

The queen. Every hive must have one. The queen is the most important member of the colony because she is the only female that breeds and lays eggs. She ensures the future of the entire hive. Through her chemistry, she also controls the hive and tells the workers what to do.

Workers. An all-female force of many roles. Workers are guards, babysitters, farmers, and insurance agents. They are the ones that collect pollen, build combs, raise bee babies, and protect the queen from danger. They make up the majority of the hive community and they'll die to defend it. Worker bees run their shop like type-A personalities with obsessive-compulsive disorder — they are ridiculously clean and hardworking. If a worker or drone dies on the job, they haul it out and take it away. Workers won't even suffer the inconvenience of a grieving time. These girls mean business.

Drones. Drones are males, and they don't really do anything but fly and crawl around hoping they'll get to have sex someday. Drones live to mate with the queen and then die shortly thereafter (usually because the workers have kicked them out into the cold). They are needed to breed, but that's pretty much it. Which makes them both the most vital and the least employed members of the hive. Men.

BEE STUFF

Chances are, you don't have a friendly neighborhood bee source, so mail order is the way to go. (Turn to Research, Son, page 188, for a listing of beekeeping supply companies.) Bees come by the pound, and the general starter weight is three pounds with a queen. If you have the option, order your queen "marked." This means that even if you don't know your bee classes by sight, you'll know the queen by the big white dot painted on her back. You'll need to order your own starter colony well in advance for the spring start (most suppliers are out of bees by February, so place your order as soon as your apiary starts taking them).

Hive body and frames. You'll need a basic hive body with removable frames and starter combs. Starter combs are false combs used to give the bees someplace to begin their building process. Consider yourself a contractor laying down a foundation for the construction workers. These combs may be either plastic (perfumed to smell like beeswax) or real wax. (If you plan on harvesting full comb squares, opt for the pricier edible wax.) Also, when ordering that original hive body kit, make sure to include a feeder. This is just a glorified upside-down Mason jar that you'll fill with sugar water. It's a standard piece in most sets, but just in case, be sure you specify it on your order. During times of peak honey production, you'll also need to add a "super" — a hive box that the bees use to store reserves of pollen and honey.

Hive tool. This is the all-purpose beekeeping tool. It looks like a mini crowbar and is used to pry apart combs, take out nails, and cut through wire. You should always have it with you when you go out to the hive.

Smoker. A smoker is a metal can that pumps smoke into a hive. Some beekeepers light dead leaves and bailing twine inside the smoker; others use a special compound sold especially for smokers, which lasts longer. The smoke is used to calm the guard bees that protect the hive, so they won't attack you when you come to work in the hive.

Protective attire. You don't need a full-on beekeeping suit; any light-colored jacket and sweatpants will do the trick. But I advise you to purchase a hat, a veil, and a pair of good bee gloves to protect your most vital beekeeping tools: your hands and eyes. The hat is wide-brimmed, so that the veil hangs well away from your face and neck. The gloves are plastic and have long sleeves that reach almost to your elbows. Since your hands and head are the most vulnerable to stings from wayward bees, it's best to keep them covered until you get truly comfortable working with the hive.

SETTING UP THE HIVE

Carefully consider where you'll locate your hive before you set it up. It should be in a place that isn't beaten by the sun or vulnerable to heavy rains. But you also don't want it to be so sheltered that it's at risk of being knocked down by falling branches or that proper sunshine can't get to it. Common sense says to put the hive in a place where you'd want to pitch a tent if you were staying for a whole year.

Many beekeepers recommend setting up the hive to face east, so that it gets morning sun and triggers early foraging. It's a good idea to locate it with a tree nearby, in case the bees decide to swarm (a fairly common occurrence). If there's a tree near the hive, a swarm will likely settle on a branch, making it easier to capture and relocate back to the hive.

Once you've found a site for your girls, use a couple of cinder blocks or an old table to elevate the hive body at least two and a half feet off the ground, to keep mice, raccoons, annoyingly brave four-year-olds, and other random nuisances from wreaking havoc. Bees also hate being dirty, so elevating the hive will help keep out mud and floodwaters. If you're in bear country, consider setting up an electric fence around the hive.

One last thing: It helps to locate the hive within view of the house. If there's a spot you can see from a window you pass often, you're more likely to glance out and notice when the lid is askew or if the feeder fell from the shelf during the night. Seeing little things like that could make all the difference to the bees' survival that first year.

. . .

TENDING BEES

Good news, friends. Bees pretty much tend themselves. At least, they're not like other livestock, which need daily feeding and watering. In fact, they might be the only farm animals you can safely leave unattended while you go on a monthlong tour of Romania. But there are a few things you should make a habit of doing every so often. Be sure to open the lid of the hive at least once a month to check for signs of disease, mold, population ballparks, and general health of the colony. Along the way, you'll need to be on the lookout for such things as American foulbrood disease and varroa mites. See Research, Son, on page 188 for a list of resources that can help you identify and treat pests and diseases.

Also, make sure water is available to them if there aren't any natural sources nearby. Some beekeepers get the watering fonts used for chickens, prop them up on a pedestal near the hive, and change them every week or so. If you're uncertain about the availability of pollen or nectar (especially true for rooftop hives in urban areas), you can provide your own pollen substitute and sugar water to help them out along the way.

HARVESTING HONEY

Once harvesting honey is possible, the process is quite simple. You can either slice the combs from the frame with a comb-slicing tool or you can extract the honey. Harvesting the combs and honey in one piece is simpler (not to mention less expensive — even the cheapest hand-cranked extractors cost well over a hundred dollars), but it means your bees will have to spend precious honey-producing time rebuilding the comb next season. If those golden jars are what you crave, you'll need an extractor. This cylindrical tool holds the frames in its center and spins fast and furiously, whipping the honey out of the combs and onto the walls of the extractor, letting the sweet goo slide to the bottom for collecting in jars and bottles.

A friendly warning, though: If there's one time during the whole beekeeping endeavor when you're in danger of getting stung, honey stealing is it. So suit up in a few layers and wear as much protective clothing as possible. Tuck your pant legs into your boots and make sure your veil is secure. Give the hive a hearty dose of smoke before you pull out the frames, and work quickly.

JOIN A LOCAL KEEPER CLUB

If beekeeping seems a little too daunting to start on your own (I completely understand), consider joining a local apiary society. Most states have city and regional beekeeping clubs that offer classes. These clubs aren't just for education, either; they're an entire community at your fingertips, with phone numbers and enough collective experience to make almost any emergency manageable.

THE COUNTRY KITCHEN

Bringing the heart of the farmhouse into your own home

THE FIRST TIME I held a perfect loaf of butter-top browned white bread that I'd made myself, I felt like I had just finished an eight-mile summit; it was just as rewarding and emotionally exhausting. There is some magic to it, too — the chemistry of the yeast and the brushing of melted butter over the soft dough before it goes into the oven. When it's finally baked, you can't decide if you want to chomp into it or wrap it up in paper and string and give it to the first person you make eye contact with. It's a great problem to have.

I remember holding my first loaf in awe in the winter sunlight of my Idaho kitchen. The dogs were at my feet, looking up at me with their tails wagging, hoping for an impromptu snack. But the idea of ripping into it seemed almost sacrilegious for a few holy minutes. So I stared at it like a crazy person until Annie had enough and jumped at it, teeth gnashing. Which brought me down to earth and taught me a lesson — stop acting like you did something special and dig in, lady. I set the warm bread on a plate out of the dogs' range of theft and cut off a steaming slice. I spread on some butter and watched the edges melt into the little pores. The honesty of that first pristine country loaf made the world seem a little simpler. Outside was a cold, ice-covered place, but in my palms was a warm oasis.

SOON I WAS ABLE to share my paradise with friends. One evening in July my newlywed friends Sara and Tim called to say that they would be spending a few weeks in Seattle, and would it be okay if they drove over to Sandpoint for a weekend to visit? Okay?! It was more than okay — I was ecstatic. I missed them, along with all my friends back east, and knowing that someone I once shared a dorm with would soon be sharing a slice of quiche right here on the back deck was staggeringly wonderful. When you up and move to Idaho, it's not easy to get people from Pennsylvania to come hang out with you — with jobs and families and lives so regimented by obligations and manager-approved vacation days, it's hard to get the old gang together for beer and pizza. Especially when half a continent is between you . . .

But Sara and Tim were pulling it off. Sara was on summer vacation from her job as an elementary school art teacher, Tim had just finished taking his entrance exams for med school, and both of them wanted to relax. After all the months of screaming seven-year-olds and endless studying, they wanted their short visit to be filled with no plans at all except chilling out and seeing some mountains. Perfect. Far as I'm concerned, the best kind of guests are the ones who don't want to do anything but lie on the grass and swat at flies. The visit would be epic in its inactivity.

Or at least it would be for *them.* While Sara and Tim had no expectations for me, I had pretty high expectations for the farm and the weekend in general. I wanted them to feel like this place was special, because it was. Guests were a rarity here, and this was the first time in my life I could feed people from my own garden, so my inner B&B hostess came out. I mowed the lawn, lined the chicken coop with fresh hay, simonized the rabbit cages, bathed and brushed the dogs. Hell, I even dusted.

After I finished the housekeeping, I planned out a menu, baked fresh bread and cinnamon rolls, and prepared a quiche for the fridge. I wanted to treat them to the delicious foods I had been enjoying the past few weeks. I wanted them to understand what all the hard work was for. I also wanted a witness. There is a special kind of validation in the smiles of close friends, and part of me needed to see that so I could remember it when I was wading through three feet of snow to feed the chickens once winter set in. I was certain that happy July weekends would make February snow easier to trudge through.

When Sara and Tim finally did arrive, the weekend went brilliantly. We read books in the sun and drove aimlessly through the mountains. We had campfires in the yard and held rabbits in our laps. Every meal we ate came from my own birds, baked goods, and soil. For breakfast we ate omelets and quiche. For lunch we buttered freshly baked breads and tossed salads. Dinner was a potato and pea soup we invented on the spot or homemade pasta with fresh garlic bread.

With Sara and Tim sharing meals by my side, I realized everything I had been sweating and planning and dreaming for since

I started growing seedlings back in January. When your friends fall asleep on the couch with full stomachs from the food you've produced and cooked for them, it gratifies you in a way your expectations could never have prepared you for. When you can give people that sense of fullness and contentment, you start seeing your garden and kitchen and chicken coop in a different light. And I'll say this much with complete certainty — it's a hell of a view.

RISE UP AND BAKE

Baking bread, stovetop canning, and making good coffee represent the three holy Ps of my farmstead kitchen – producing, preserving, and percolating. Try starting with a fresh loaf of bread. The additional recipes starting on page 162 will provide you with homemade food for every meal of the day. They're also easy to learn and rewarding to share.

MY HOMESTEAD CABINET

When I began experimenting in the kitchen, I decided to get my supplies from secondhand sources, from online stores to local thrift shops. After all, there's no reason to spend extra money when you don't have to. Some people get a little squeamish about using secondhand pots and pans, but there's no need for that. Cooking isn't neurosurgery. Just clean your finds before you use them, and you'll be all set.

Here are some basic supplies you can use to transform your kitchen into something even the original pioneers would deem respectable. (Actually, I think if the original pioneers were in our kitchens, they'd be terrified. And possibly convinced that our microwaves were direct portals to the underworld.)

. . .

Cast-iron cookware. Nothing beats a cast-iron skillet for country cooking — these babies are dark, heavy, and feel great in your hands. You can now get them "seasoned," which is a fancy way of saying already prepped for cooking. If your skillet isn't seasoned, give it a good scrubbing, rub it down with cooking oil, and set it in a low oven for a few hours, to create a protective coating. I have a small skillet for pancakes and a big one for large batches of eggs and for stir-fry.

Large wooden spoons. Large wooden spoons are another homestead necessity. I use them for everything, from mixing batter and stirring stew to shaking at meddlesome neighborhood children. Wooden spoons are ideal because, unlike their metal brethren, they don't burn your hand off when they've been sitting in a pot too long, and they won't scratch the bottom of a pot.

Hand-powered tools. Besides not using unnecessary electricity, hand-powered tools give you more control and slow down the process enough that you can relax while you're in the kitchen. All of my mixers, grinders, juicers, and processors are hand powered.

THE BEST COFFEE EVER

Locally roasted whole beans
Organic sugar, to taste
Organic milk or soy milk, to taste
1 bar of dark chocolate
(I'm serious — do *not* get milk chocolate; we're all adults here)
Coffee grinder (mine is hand-cranked)
Metal percolator (electric or stovetop) with metal basket

I love coffee. I love coffee so much, it's almost indecent. I take more pride in it than I do in the most meticulous pie recipe. Good coffee means fresh whole beans you don't grind until right before you turn on the percolator. It means pure water and real cream and the best sugar. But none of that matters if you're not percolating it. Forget modern plastic coffee makers, French presses, and espresso machines. If you want amazing coffee, find an ugly metal percolator made before 1978. Clean it up, replace the cord if you need to, and use it with love. That's a real cup of coffee. I didn't realize this until I bought my first percolator out of a junk bin at an antiques mall. I think I paid seven dollars for it. Three states and endless mugs later, it still makes hot, delicious coffee every morning.

Fill the percolator with water to the level indicated on the inside of the container (between 6 and 10 cups). Grind beans and fill the basket until your index finger is in ground coffee up to the middle knuckle. Plug in the percolator, or set it on the stove to brew if it's the old-fashioned camp style. Break into pieces as much dark chocolate as you think you can handle (I go for half a bar when I'm feeling frisky) and put them in a large mug. When the coffee has finished percolating (5 to 10 minutes, depending on how strong you like it), pour it into the mug and add sugar and cream as desired. If the coffee is too dark for you, add a bit of hot water.

Remember that coffee can be ruined when it's left too long over any kind of heat. When you've had your fill, turn off the percolator or pour the coffee into a thermos or carafe.

BASIC BREAD

Makes two loaves

2 cups warm water
1 packet active dry yeast
1 teaspoon honey
2 tablespoons vegetable oil
1 teaspoon salt
5 to 6 cups unbleached white flour
Butter

I make bread on Sundays — it's so nice to end the weekend with the comforting aroma of baking bread! Usually I like to make a pan loaf to slice up for toast and sandwiches and a nice braided loaf for tearing off piece by piece for snacks and dipping in soup. Leftovers get reincarnated as French toast. This basic recipe can be used to make everything from garlic sticks to cinnamon rolls. Since the dough takes some time to rise (about three hours total, even with fast-acting yeast), it's great to plan your baking around errands or laundry and other chores.

Prepare the yeast. To begin, you need to get the yeast started. In my experience, this is the trickiest part of the whole endeavor. Pour 2 cups of warm water — around 100°F, whatever you can dip a hand into without real discomfort, but that's clearly above room temperature — into a large bowl.

Empty a packet of yeast into the water and stir it until it dissolves. Once the little grains of yeast have disappeared into a foggy, tan murk, add a teaspoon of honey and mix that in as well. Let it sit for 5 to 10 minutes. After it's been given its quiet time, you'll notice bubbling. This is a good thing — it means you're activating the yeast, bringing it to life Mary Shelley–style. Frothy yeast water means you pulled it off. Congratulations.

Start the dough. Mix the vegetable oil, a teaspoon or so of salt, and 1 or 2 cups of the flour in with the yeast. You really need to beat it together. If you have a hand mixer, go to town. And if you're using a trusty wooden spoon, keep at it for 3 to 5 minutes. If you don't think baking can be a workout, opt for the spoon and watch your biceps grow.

Add a cup of flour at a time, not more than three additional cups. Now, once you have a ball of dough that isn't too sticky to handle and has some elasticity, you're ready to knead. Lightly flour the (clean) countertop and turn out the dough onto it.

Knead the dough. Kneading dough is cheap therapy. Punch the dough and roll it into a nice mound. When you think it's been kneaded enough, knead it some more. There's something called the "baker's window": that's when the dough's been massaged so deeply, you can stretch it into a membrane so thin that you can almost see through it. That's when you know you've done enough. You can also tell it's ready when you poke it with your finger and the dough springs back easily, without leaving a dimple.

Clean the bowl, grease it with some butter, and place the dough in it. Give the dough a spin to give it a layer of butter and then flip it over. Cover it with a clean damp cloth. It needs to rise in a nice calm place for about 1½ hours, so if you don't want to sit on a stool and stare at it, find something else to do. Run to the post office, throw in some laundry, call up an old friend. When you come back to the covered bowl, you'll know it's ready when it's doubled in size.

Now punch down the dough. Really. Go ahead, make a fist and punch the trapped air out of the dough. When you've pressed out all the air bubbles, put the dough on the counter and knead it a little more. When it looks like it's back to that original consistency, rip or cut your dough in half. Shape half into a loaf and place it in a greased bread pan. Cover it with a damp cloth and set it aside. Divide the remaining half into three parts. Roll each part into a chubby snake and connect all the snakes at one end. Make a simple braid, press the ends together, and tuck them under tight, so it doesn't unravel while it bakes. Set the braid on a greased baking sheet and cover it with a damp cloth. Let both loaves sit for another hour, or until they're doubled in size. Put on a favorite movie or find out what's on the radio. Better still, check out some bread recipes or hunt down tips online.

Baking. When the dough has risen, you're finally ready to bake. Preheat the oven to 375°F. If you like, brush melted butter on the tops of the loaves. This adds a little color and lots of flavor. You can add herbs or garlic salt to the melted butter, if you like.

Bake the loaves for 25 to 30 minutes. You'll know they're ready when the dough is browned and a little hard at the top. A loaf will also sound hollow when you thump it. If you want to be extra sure they're done, stab them with a sharp knife and see if it comes out clean.

When you pull out the loaf pan, let it sit for a few minutes, then take the bread out of the pan. You can do this easily if you hold the bottom of the pan with a pot holder and have a kitchen towel guarding your other hand from bread burn. Turn the pan upside down and let the loaf slide out gently. Use a spatula to remove the braid from the baking sheet. Set both loaves on a rack to cool. Let them cool for at least an hour, then slice and serve with butter.

HOME-CHURNED BUTTER

Makes about ½ cup

Carton (1 cup) of heavy cream
A dash of salt
Pint jar with lid

Making your own butter doesn't require a giant churn or a cow. It's just whipping heavy cream in a jar! Buy your cream from a local dairy farmer, if you can.

Fill the jar halfway with cream and let it sit out overnight (8 to 10 hours). This lets the cream set and cure at room temperature. The next morning, start shaking it (not crazy, no need to be violent, just a simple shake once a second). After about 15 minutes of shaking, you'll see the buttermilk start to separate from the butter, which will begin to form a pale yellow ball. Pour out the buttermilk and save it for making pancakes or biscuits. Knead the butter under cold running water for several minutes to incorporate the fat and to work out any leftover milk (which will make the butter go rancid quickly). Then knead in the salt and store the butter in a covered container in the fridge. As you get hooked on churning, you might want to order wooden or metal butter molds, to make your butter be a certain shape.

You can make the churning process go faster by starting with very cold (rather than room temperature) cream, and by adding a cold marble to the cream before you start shaking.

THE KITCHEN RADIO

I can't imagine a kitchen without a radio. There is something intimate about live radio that television programs and audio books just can't re-create. Listening to a story, news, music, or even the local weather report unaccompanied by visual distraction lets you give the majority of your attention to the task at hand while staying in touch with the world in a way that people are slowly forgetting. Another often overlooked perk about radio is you don't need to pay a monthly invisible-wave bill. Once you get that magical box, all the great stuff that comes out of it is free. Some radio shows, like Chicago Public Radio's *This American Life,* have become staples of my weekly routine. I can't imagine a Thanksgiving without tuning in for its annual Poultry Slam.

OLD STUFF

Antiques for the renter's homestead

I NEEDED A CHEESE GRATER. Not the most stimulating of needs, but a need nonetheless. I was in the middle of a recipe that called for grated cheese, and I refuse to buy the shredded bagged stuff. In my book, preshredded cheese is the equivalent of taking the elevator up one flight of stairs — it's just lazy. Anyone who isn't wearing a pair of arm casts and has three minutes can shred a block of cheese. Anyway, I hadn't planned ahead. I was making a quiche, and in the excitement of baking with my own hens' eggs for the first time, I hadn't thought through the whole giant block of Cheddar on the table. I wasn't about to dice it, and I wasn't going to bake it like a wheel of Brie, so in lieu of going out to get a bag of cheese, I headed to the local antiques shop.

I know there are a lot of ways to obtain a cheese grater. I could go to the kitchen-supply store. I could check the home-goods section of the supermarket. I could buy one made from recycled soda bottles online and have it shipped to the farm. But none of those would do. When I need something, I go for the old stuff. I wanted an old metal grater that had been around for a couple of world wars and was still going strong. Perhaps one with a faded-paint wooden handle, or maybe a postwar bright teal job from the fifties.

After about an hour of hunting around two large secondhand shops, I made my kill. Hanging from a nail on a support beam in the basement of a shop was my prey: a long metal grater with little holes in weird triangle designs. I had no idea how old it was, but my guess was that it was from the forties. It certainly was well used and lacked the sparkle and surreal futurism of the atomic kitchens of the fifties, the plastic of the sixties and seventies, and looked a little too shiny to have made it through the Depression. Regardless of my guessing, it cost four dollars, looked cool, and could stick to the fridge with a strong magnet. Score.

Antiquing for everyday household items isn't exactly a common practice. A lot of people are out there eyeing the shelves of flea markets and junk stores for the rare collectible or nostalgic item for their bar walls, but I go for the things I use every day. I go for coffeepots and toasters and spatulas. I go for tables and lanterns and records I want to listen to. At work I use a Fire-King Jadeite mug from the fifties because I'm positive coffee tastes better in the green milk glass than in any other container in the world (try it and you'll never go back to ceramics).

There are a lot of really good reasons I run to the past when I need something as utilitarian as a cheese grater: Things were made better, looked prettier, and lasted longer before plastic took over. Buying from neighborhood secondhand shops helps support the local economy and is a kind of recycling. But I also go vintage because I want my home to be full of awkward, lovable items that are ingrained with memories of people and stories I could never forget.

MANY OF THOSE STORIES come from the years I spent antiquing with my closest friends in college. For four years I went to art school amid the cow pastures and Mennonite farms of Kutztown, Pennsylvania, at the local university of the same name. Kutztown was and still is a sleepy college town of trains, farms, gray cemeteries, and black buggies. Even with the occasional fraternity front porch, it's still a real charmer. There's a small downtown district with rows of turn-of-the-century storefronts that contain coffeehouses, cafés, and gift shops. If you're jonesing for some feline attention, most of the local establishments have a cat sleeping on a cash register or in the front window. (It's pretty much all the quaintness you can ask for an hour outside Philadelphia.)

Like a lot of creative kids, I wasn't that into sports. I was a member of the equestrian team for a year because I wanted to learn to ride and wanted to be around animals, but it never was a "sport" to me. Riding was more about learning to understand the animals and getting them to understand me. I realized I didn't have the desire, cash, or energy to be of use to the highly competitive team, so I backed out. I didn't miss the meets, but I did miss the camaraderie. Getting up early in the morning and traveling around the tristate area with friends was just plain fun.

A few months later, though, another team formed and I was a charter member. A ragged group of us started a weekend tradition of early-morning excursions. While it had nothing to do with horses, it had a lot to do with being part of something special.

I had a college sport again: it was antiquing. I nicknamed our group the Order of the Peculiar Materialists.

NOT TOO LONG AGO my friend Kevin called to tell me about a new antiques store that opened in Kutztown. He described it as simply "not having our stuff." I knew instantly what he meant. When I got off the phone, I couldn't stop thinking about "our stuff." Somehow during the years of aggressive group antiquing, we created a library of items that became a part of our own brand. It was a peculiar mix of things any of us would pick up: West Virginia glass, Jadeite mugs, Eames and Knoll furniture, Maruska prints, retro kitchen appliances, Lincoln Beautyware, typewriters, Pyrex bowls . . . The list goes on and on. I don't know if these things actually went together or if we just decided they did. It didn't matter.

We were a rogue legion on Saturday-morning excursions. We got up early; usually by nine a.m. we had already swapped snide remarks about overpriced flour canisters and had a cup of overly sugared coffee or Earl Grey in our hands. While other college kids were fighting hangovers or just happily asleep, we were haggling over the price of old plaid thermoses. But this was what we wanted. This was our sport. We played home games at Renninger's (a giant indoor marketplace) and other local arenas. Sometimes we would take group or paired trips to away games in surrounding towns. Every season, all types of weather, all kinds of budgets, we were cutting a slim berth on manic and loved every second of it.

We trusted the team's taste and soon learned each other's. Each of us had a sensibility or a period that really seemed to suit us. I don't know if I had a period — I liked suitcases, taxidermy, fraternal-order collectibles, teal kitchen appliances, and hard-faced stuffed animals. Anything else I collected was because it reminded me of the group or someone in it. I liked that a lot about my stuff. I like it even more now that I'm so far away. I can still pick up any item in my house and remember the events, people, and conversations behind it. It's my intimate museum of strangers' objects.

I don't know what the others' motivation was: maybe they were nostalgic like me, or maybe they just liked cool, weird things. Regardless of why we collected, the effects it had in tightening the core group were undeniable. Not only did these people become close friends; they were decorating with things we went out of our way to get together. By default or desire, the act of obtaining those things meant more time with those people. Everything on the walls held the dialogue that was going on when it was first picked up. Every appliance in the kitchen was purchased with the nod of a fellow Order member back at the cabal. As soon as you walked into another Peculiar Materialist's place, you were surrounded by the memories of countless weekends. It was a real testament, and you could reach out and grab it off the shelves.

We remembered. That's the entertainment center you made into a terrarium that night on the porch with Bob. It was summer; a whole cast was there. There were fireflies and cigarette smoke and cheap glass votive holders with rainwater in them. There's the TV you got from the pop-art store in Bethlehem for fifty bucks. We couldn't believe it worked, and we carried it home in my old Jetta. The sun was so obnoxious that afternoon. Right in my eyes. On the way home, I was scared we would get in an accident. You're wearing the necklace you found on the rack in

Jim Thorpe. You got it from the place I ran back to, so we could all get Kevin that Civil War jacket for his birthday. My god, Jim Thorpe.

After I left Kutztown, it took months to want to start seriously devoting time to this again. There was no one to go with. This was a team sport, right? No one I knew understood what was absurd or wonderful. No one I met wanted to get in on the joke or to share in the stupid glory of finding an item we all appreciated. When we graduated and moved away, the Order of the Peculiar Materialists was disbanded.

However, a few months later the urge came back. I wanted to find an original Moonbeam alarm clock. The idea of coming across one in working order sparked that same old excitement. No one from the old Order was going to see it when I stumbled across it. It would be a silent victory.

STOCK UP ON OLD STUFF

Why go vintage, anyway? The simple answer is that things were made better back when people didn't plan on replacing them every ten months. Stuff just doesn't last anymore — new things break at the slightest inconvenience to their molded-plastic structure. The way goods are manufactured now, you have to pay ten times the price to find something as reliable as your standard antique.

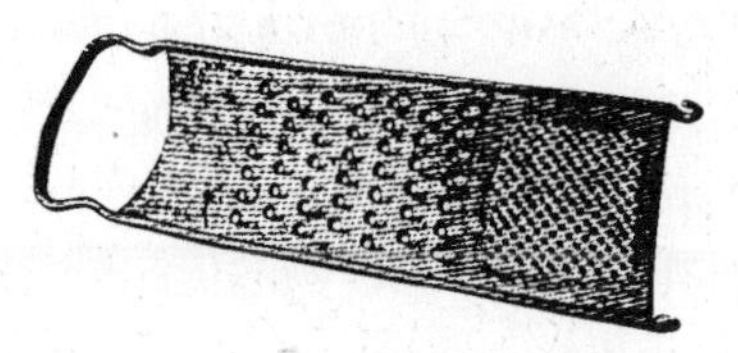

A GREAT EXAMPLE is an electric fan. When I moved into my college dorm, I bought a ten-dollar desk fan to cool me off. While I was studying, it got pushed over by some books and fell four feet onto carpet. The plastic blades broke. Had I driven three miles up the road to the antiques dealer and bought an old metal Westinghouse fan for fifteen dollars, I wouldn't have had that problem. For one, it would be too stout and heavy to be easily knocked over by a few paperbacks, and if it did somehow fall to the ground, it wouldn't crack in half. If something ever did break, it could be fixed.

Besides being better quality than whatever's lining the shelves of big-box stores, a lot of vintage wares that are in good working order are easier to repair yourself. They come from a time when things were made with integrity, and when they stopped working, anyone with a basic tool kit could whip out a screwdriver and figure out how to fix them. Tighten a bolt here, twist a wire there, and bam! You've got music coming out of the radio again.

The benefits of going vintage aren't just practical — the stuff is beautiful. Design from the fifties through the seventies is amazing. Everything from old lamps to lounge chairs has a utilitarian simplicity that's lacking in modern designs. Only the most current designers are returning to what worked back then. If the modern look isn't your thing, there are plenty of older themes. There's the Art Deco beauty of the twenties, with flappers on ashtrays and sleek cats on clocks. There's the indisputable practicality of the forties, with milk-glass mixing bowls and metal lunch pails. And there's the elaborate intricacy of styles going back even further than that. Think Victorian or Revolutionary. Or go in another direction and collect folk art from the backwoods, or the rustic creations of loggers and miners out west. It's only pricey if you're looking in pricey places. Not everything old is junk or a treasure. Most of it is just plain old stuff, still lying around and completely useful. So instead of wasting money, packaging, time, and energy, buy old.

RECYCLING NEVER LOOKED SO GOOD

Recently, my favorite percolator croaked. I'd repaired it three times, and it was just its time to go. I was its second owner in more than forty years of nonstop service, so I can't really complain. I was in the market for a new old one, and that familiar junk-hunting sparkle returned to my eye. I had a specific need to fill and it was the perfect coffeepot. The requirements were as follows:

- It had to be all metal (metal lid, metal basket, metal everything but the handle).
- It had to have the little glass bubble in the lid for perking goodness.
- It had to look totally different from any newfangled coffeemaker on the current market.
- It had to cost less than twenty dollars, shipping included.

Thanks to the wonder that is the Internet, I found a replacement on eBay in twenty minutes. It was a metal contraption from the fifties, shaped like a giant teapot with a green wooden handle. Right on. It was in perfect working order, and with shipping it came to only fifteen bones. The cheapest ten-cup coffeemaker at Target was twenty-nine dollars, and it was plastic. The modern options on its site looked like every other coffeemaker in break rooms across America. But this old thing, it had moxie. I had been scoping out old stuff for years and never saw anything like it. Kismet smiled down, and within days I had an awesome new caffeine enabler delivered to my front door. Someone else's burden became spare cash and my trophy. Welcome to the coolest way to recycle since vintage consignment shops started restocking for hipsters and hippies alike.

Every time you go out and buy a set of glasses at a yard sale instead of ordering a new set online, you're taking part in recycling. In a renter's world of apartments furnished by Ikea and Walmart, be intrepid and go retro. Stand out from the crowd and hang a cuckoo clock on the wall or throw an old bearskin rug on the living room floor. The world is already chock-full of all the cool furniture, place settings, and sport coats we could ever need, so why keep adding to the excess by buying new ones? You'll have more fun finding the perfect items to suit your specific taste, and you'll probably save a ton of cash doing it.

OLD THINGS I SUGGEST

If you're new to this concept, here are some recommendations on what to look for when you first venture out into the antiques malls. If you're not familiar with the things listed below, you can look through collector's guides at the library or do a simple Internet search for visual aids, but chances are, you've seen these guys around. After a few weekends out, you'll get the hang of it. The basic rule to follow is, when you need something, consider buying old first. Before you know it, you'll be calling your local dealers when you need a metal cheese grater mid-recipe.

Glass dishware. This here is the real deal, folks. It's chunky, heavy, and it was microwavable before there even were microwaves. Fire-King is my favorite kind of glass dishware — it's top shelf. Although these dishes have become collector's items, you can still find boxfuls at garage sales and auctions. Another perk is that these mugs and bowls come in so many patterns and colors. There are polka dots, animals, stripes, and scrimshaw on these babies. Your limits are set only by how long you're willing to hunt for them.

Pyrex bowls and baking dishes are the hefty ones your grandmother and mom had around the house (and probably still do). I've started collecting a blue set that's decorated with Amish couples and chickens. I use them every day and wouldn't know what to do in the kitchen without them.

Hand-powered kitchen appliances. If you've got the upper-body strength, hand power is the way to go. You'll save energy, be able to grind fresh coffee beans in a blackout, and get to burn a few extra calories in the process. Start with mixers, grinders, churns, and food choppers.

Westinghouse fans. For fresh air on hot nights, my Westinghouse does the trick. It's more than fifty years old and works like the day it was ordered from the pages of the Sears, Roebuck catalog. Whoever designed these metal, oscillating marvels has earned a special place in my heart. I'd get a tattoo of one if I were into tattoos.

Old suitcases. For storage and stacking, I love old boxy suitcases. I've made bookcases, coffee tables, and filing cabinets out of them. I also use them as luggage (how novel). Maybe it's just the traveler in me, but regardless of why I have them, they are great for organizing everything from first-aid supplies to magazines.

Record players and records. For the best outdoor entertainment since the Slip 'n Slide, put a 1970s record player out on the deck at your next barbecue and prepare to be swarmed. Those old turntables at the junk store still work, and everyone's got a stash of great records at home. Vinyl's still being recorded by any artist worth listening to, so there's no excuse not to have a phonograph in your apartment.

Metal canisters. For kitchen storage, Lincoln Beautyware is hands down the most charming and well made of the lot. Depending on your own aesthetics, you can find these looking like they just came out of the box or beat all to hell. I like my old stuff a little banged up. It reminds me that it has a history bigger than my own. These are for the essentials you'll be stocking in your pantry — flour, sugar, coffee. You get the idea.

Old radios. It's not hard to come across old radios that are still in fine working order. For some reason, people toss their old radios long before they're ready to meet their maker. You don't need to hunt down a 1950s teal table topper, either; you can find old clock radios from more recent decades at just about any yard sale.

If you want to invest in alternative energy, you can order a crank-powered radio that doesn't even need batteries. You generate power by turning a small lever for a few minutes, and hours of clean-energy entertainment will come streaming out like gangbusters. These are great for nights on the trail or extended hiking trips. And when the power goes out, you'll still be rocking long after the neighbors' batteries have died.

DIY WARDROBE

Clothes built by Jenna (or you)

I DON'T THINK I EVER had homemade clothes while I was growing up. If I did, I sure didn't know about it. One of the hard truths of my suburban elementary school culture was that homemade clothes were completely taboo. If you showed up for class wearing something you didn't buy at the mall, you were considered a freak or borderline homeless. Cruel, but hey, that's the way eleven-year-olds are. Even today, when I find out someone is sporting a handmade shirt, I'm a little shocked — not because of any residual elementary school judgment (I'm pretty sure I outgrew most of that), but because it never looks homemade.

Take my friend, Marjan, for example. She comes into the office wearing these amazing vintage quilted fabric and wool skirts that look like something ripped from the pages of Anthropologie's winter catalog. That is, if Anthropologie took a Marjan-specific order, taking into account all her peppery personality traits and precise measurements. Her handmade clothes aren't camp or boxy — they're perfect for her. And the coolest thing about them is that nowhere else in the world is someone else walking around in anything quite like them.

The Marjan Collection had a pretty big effect on me. Seeing something as simple as a coworker in a cool skirt was all it took for me to understand that I had been a chump in elementary school. To me, it's almost magical that someone could whip up an A-line skirt or pair of pants from a bunch of fabric and thread, and I think a lot of people feel the same way — sewing a garment is a whole different level of accomplishment from any other skill. You can take a delicious casserole to a potluck, and the host will say thanks and ask what's in it with polite interest (likely out of concern for allergies). You can show your friends the giant, juicy tomatoes you grew or the crazy chickens in your yard, or play them a few self-taught chords on a guitar, and they might nod and show mild enthusiasm. But when you show up in a pair of jeans you stitched up from a "Built by Wendy" pattern (see Research, Son, page 188), those same people's eyes widen and their voices lower in amazement.

Most of us never even consider that something like a pair of jeans could actually be made without an assembly line behind it. Like the meat on our plates, the clothes on our bodies have been far removed from the process that created them. I understand why people wouldn't want to slaughter a cow to make their own hamburgers; it's a bloody business. But buying fabric and making a T-shirt isn't exactly gruesome. So why do we run to the store every time we need a new pair of pants or a dress, rather than sew it ourselves? It's a reflex; for generations, nearly all the clothing we wear has been factory made. We've all come to expect that someone far away, probably in a country we'll never visit, will sew our clothes for us. Making them at home doesn't even cross our minds. Well, maybe every once in a while it should.

THE FIRST TIME I sewed my own piece of clothing, it ended up looking like a poorly assembled prop from one of the *Lord of the Rings* movies. No joke. I found a pattern for a hooded poncho, with a picture on the package that made it look elegant but also easy to make. It had only three sections to cut and sew together — I thought I was home free. I bought expensive fabric and got out my portable machine. It would be like trying a new recipe: as long as you followed the directions, you'd be set, right?

Once I got started, I realized that I hadn't bought enough fabric, so I had to shorten the pattern. This was beyond Sewing 101. Shortly after that, I found out the pattern didn't include directions for any sort of lining. Trying to fake my own fancy lining put me about midsemester in Sewing 501. So, after four frustrating hours spent working on a slow machine, I produced an awful little cape with bunched-up, meandering seams. It looked more like something a wet, angry hobbit would wear than a cool find from the Gap. *Sigh.*

Lesson learned. I went back to the fabric store with a humbler attitude about making clothes, but determined not to let that horrible poncho experience bring me down. Not wanting to waste any more money on Renaissance Fair costumes, I decided to try a less expensive route for my second attempt. I found some mustard plaid flannel on sale for ninety-nine cents a yard. It was ugly but cheap and warm. I picked up a pattern for pajama pants and decided to give it a go. My sewing guidebook warned that making pants could be challenging for a novice seamstress, but the picture on the package looked friendly enough and made me want to drink cocoa. Sold.

I brought my goodies home, put in a DVD, and in less time than it took to finish watching the movie, I was trying on my own comfy drawstring pj pants. I even hemmed the cuffs and used scraps to sew a cord for the drawstring. I put on my creation and did a little celebration dance for the dogs. They were less excited than I was, but I think it was just their elitist, pantless attitude that kept them from sharing in the glory of my accomplishment. The pants weren't perfect, but they were definitely pants, and they fit me in a way no store-bought pants in my price range could. So as it turned out, I had more success spending five dollars on fabric and an advanced pattern than I did meticulously executing an expensive "simple" one.

STITCHIN' TIME

Sewing and knitting are two of the most practical skills for self-reliance you can learn, wherever you live. No matter what stage of life you're in, you need clothes. Since you need to put something on anyway, why not start adding to your wardrobe a few things you made yourself? Sewing and knitting your own stuff isn't just the comeback of an old-fashioned practicality – it's a guarantee that what you're wearing is high quality, custom fitted, and one of a kind. All it takes to get started is a yard or two of fabric or a few balls of yarn.

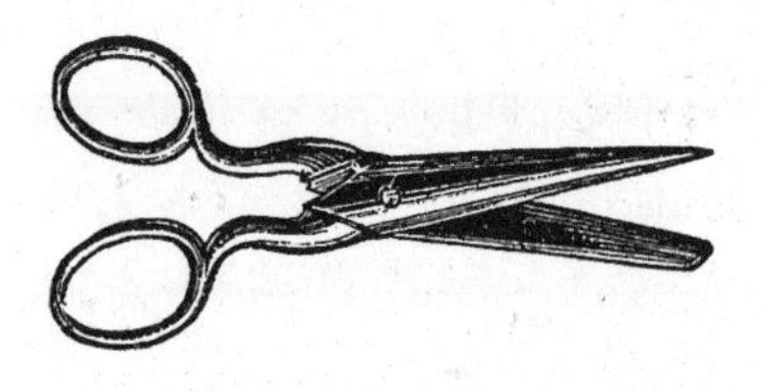

SEW YOUR OWN

Remember that amazing wool peacoat you saw in the shop window last week? Well, why not make your own in any color you want with any custom lining your heart desires? You could have that same design in olive green with a baby blue paisley lining and sushi-shaped buttons on the cuffs. Plus you'd have the satisfaction of creating your own warmth, and — short of the invention of fire (someone beat us to this) — what could be more basic to the human condition than a warm coat?

Willing to shop around online? You could find pure wool blends on sale and silks and satin prints, too. A few nights later you could walk into the diner in your own unique designs, looking better than in anything you could find at the store. And why not finish it off with a chunky knit scarf that goes all the way down to your knees? Or how about a lime green merino beanie? When you realize that you can re-create or customize anything you see in catalogs and stores, you start looking at clothes a little differently. You notice how they're put together and you take mental notes about rivets when you throw your jeans in the dryer. "So that's how they sewed on the pockets so they never fray around the edges!" Slowly, you begin to mentally deconstruct everything you wear. It's exciting when you realize that you can take apart your favorite shirt and make five others just like it — or better ones, even.

THE BARE-BONES SEWING KIT

Before you buy fabric and try to make your own corduroys, I strongly suggest putting together a basic sewing kit. This is a must-have for any would-be seamstress (except for those equatorial nudists, but they really aren't the demographic we're going for). These few items are all you'll need to start mending old clothes and sewing some basic new ones. Also, your sewing kit doesn't have to be one of those upholstered wicker baskets from the craft store. Feel free to leave those for the *Murder, She Wrote* crowd. Mine is an old blue tin ballot box with SALES REPLENISHMENT TICKETS #68601 on it in yellow writing. I have no idea what that means. I found it at a flea market and bought it because it reminded me of one of my old junk-hunting buddies. It sits on a bookshelf, like any other random antique would. However, it is suspiciously less dusty, and inside you'll find the following:

- Scissors
- Cloth tape measure
- Sewing needles
- Assortment of threads
- Straight pins and pincushion
- Seam ripper
- Needle threader
- Washable-ink marking pens and pencils
- One yard of inexpensive cotton fabric

Don't I need a sewing machine? You'll notice I didn't list a sewing machine up there, because technically you already have one (yourself). Start off by learning a few basic stitches (see Running with Stitches, page 174) and you'll be off to a good start.

Even though you really only need a needle and thread to take on even the most complicated patterns, most of us aren't hard wired with that kind of patience. Once you've tried out a few small projects (like my Easy Fleece Mittens on page 176) and know that you're hooked, you'll want to invest in a sewing machine, to save time and sanity. Consider borrowing a friend's machine or finding a cheap one on consignment. If you decide to buy a new machine, take along a friend or relative who sews. He or she will give you the lowdown about what features you actually need, while you're busy trying to decide which one looks best with your toaster. There are a lot of options out there, from computerized machines that could clear out your savings account to little guys that barely cost a Benjamin. Don't feel bad if you can't go home with an expensive apparatus that monograms towels. As long as it's in working order, the worst sewing machine in the world is still faster than you are.

Making your own clothes takes some planning ahead (making anything does, really), even if that planning takes place in the fabric store. There's no rule saying you should be able to sew together the perfect cowboy shirt overnight, so take your time. When you're looking at patterns, start off with something basic like an A-line skirt for girls or a messenger bag for guys. As with any new skill, sewing takes practice and dedication, but the feeling of wearing something you made for yourself is right up there with growing your own food. (The day I serve a meal from my garden in an outfit I made, I'll declare a national holiday.)

When it comes to fabrics, you can shop by price, texture, or department. Keep in mind the purpose of the garment you're going to sew. If you're planning on making a shirt you'll want to wear hiking, don't make it out of cotton. It takes forever to dry and wet hikers are miserable people. Beware of using dry-clean-only fabrics. It's heartbreaking to make a shawl you have to shell out ten bucks to clean when you inevitably spill coffee on it.

Most fabric stores hire staff who sew, so if you don't have an experienced sewing buddy at your side, don't be afraid to hunt down salespeople and ask for help. Tell them you're making a hooded sweatshirt and want to know if boiled wool is even a possibility and where the easiest beginner patterns are. They'll be all about getting you started.

When you finally get your fabric home, do yourself a favor and don't plan to get it all done in one day. That's like telling a beginning swimmer to lap up the English Channel just for kicks. Start by cutting out the pattern and sewing up a seam or two that first day. You made it this far in your life without sewing your own clothes, so don't burn yourself out trying to make your own spring line the same day you buy a machine.

SOMETIMES IT'S OKAY TO USE NEEDLES

Knitting takes a much smaller investment than sewing clothes does. If you want to dive into this warm, addictive world, I suggest you start knitting with a large pair of needles (around size 13) and thick, chunky yarn. Starting big is good for two reasons. One: It's easier to see what you're doing when everything is larger. Two: The results seem to pour off your needles super fast, which is important to beginners. They want to know they're actually creating something. (I've made wool hats in the time it took to watch all the special features on a DVD.) You start to see all the possibilities when you realize that every sweater starts with just a few simple rows, and you've already got that part down.

Find some enablers and get hooked. If you like working with fibers, you're not alone. There are knitting groups everywhere from uptown Manhattan to your small-town community center, and I assure you they'll welcome you with open arms. Your age, gender, and experience level have nothing to do with it.

When you do get in touch with them, be prepared to hear about everything from the Romney roving they bought online to the alpaca they just bought so they can spin homegrown wool. While not every knitting group will have online shoppers and keepers of fiber-producing animals, it's not uncommon. These are people with a passionate interest in a craft that can be as utilitarian and basic or as intricate and complicated as you'll allow yourself to make it.

For most people, learning from a real live person is usually the best way to go, but if you're itching to get started this very second, check out the books I've suggested on page 188.

WORKING HOUSE DOGS

The poor man's pony

IT WAS SO COLD last February that most of the once-soft snow froze solid. A two-inch sheet of ice covered the hay fields behind the farm, and the snow globe outside my window changed from a German fairy tale to a biting Yukon landscape. One morning I saw a brave doe try to bound across it and slide a little, even with her sharp stiletto heels. The snow wasn't deep, just slick, and unless it had snowshoes with inch-long crampons, I had no idea how any animal was getting around out there.

We humans weren't much better off. The roads in town were just as bad as my iced-over fields, and driving to and from work was starting to make telecommuting seem like the most reasonable indoor invention since the litter box. Alas, I was no telecommuter. The farmhouse didn't even have a landline phone hooked up, so I had to brave the weather like most of the working world. I bundled up, scraped the car windows, and drove the five short miles to work. Everyone in the office was sporting a giant parka and boots up to the knees. If I hadn't already known I was in the design department of a large corporation, I would have thought we were all going seal hunting.

I sat down at my desk and got to work, but I couldn't focus. I desperately wished I could be back at the farm with the dogs, watching the weather turn. The forecasters were calling for a snowstorm to come in and coat North Idaho with a few inches later that afternoon. In between meetings, I'd stop by the windows on my way to the watercooler and stare out at the mountain. The sky was heavy and overcast. Pure excitement.

I love snow. It makes sweaters warmer and coffee tastier. It makes homes feel cozier. It makes a book and a fireplace seem like everything a reasonable human being could ever need. My reasons for wanting snow tonight had nothing to do with hibernation, though. I was hoping for a quick inch to touch down and cover the ice on the road behind the house. I wanted to be out in the snow before it froze. I'm not crazy. I just live with sled dogs, and we had an appointment.

By three in the afternoon, it was almost dark (Sandpoint has the pleasure of being not only at the northern tip of the state, but also as far east in the Pacific time zone as possible; daylight is a fair-weather friend at best), and I was getting excited. I peered out the windows again: the lights in the parking lot illuminated a steady snowfall that was already covering the cars with a healthy layer. People groaned and grumbled around me. I thought, "You're in North Idaho . . . how could this possibly surprise you?" and clicked away at my mouse, trying to finish up so I could get home as soon as possible.

Driving home from work, my tires slid out from under me, trying to show me the best way to go about getting off the road. Giant plows passed me by, like a steamship passing a tugboat. I gripped the steering wheel and forged ahead. Truth is, even though the road was a horror, I was excited that it was being covered up so quickly. What makes awful traction for cars just so happens to make wonderful conditions for mushing. The fresh powder was all the traction my Siberians' paws would need to get across the ice, and the runners of my kicksled would sink onto the layer of ice below, sliding effortlessly like pucks on a shuffleboard.

One of the real perks of moving here was this weather. I grew up dreaming about dogsledding. I wrote short stories about kids who took their teams to school and kept them in a barn with a woodstove during classes, harnessing them up and taking them down the lantern-lit trails back to their family farms after school got out. I romanticized the stories of Jack London. And every now and then, even as an adult when I push a shopping cart around the frozen-food section of the grocery store, I pretend the handlebar is the grip of a wooden sled, ahead of me a team of dogs.

I FINALLY PULLED into the driveway, past the big birch tree and the rusted green truck from the 1940s. After running inside and greeting the dogs with the usual hugs and ear scratches, I dashed into the bedroom to change into heavy pants and a few thermal layers. I grabbed my red parka, fingerless mittens, and musher's hat — an old, hideous leather hat with earflaps lined with rabbit fur. (There's nothing warmer.)

Ready to go, I called the dogs into the garage and harnessed them up to our humble bispecies transport. Our ride was a small Norwegian kicksled — basically, a glorified snow scooter. It has a basket (the part of the sled where you stow gear, cargo, or a passenger), handlebars, and six-foot-long runners, but weighs a mere twenty pounds and folds flat for storage. The perfect portable dogsled for our small team. These Nordic wonders work on snow just like a wheeled scooter does on concrete. The rider gives a kick and, on level ground, it slides along for a yard or two before it needs another push. With just two dogs, it's perfect. We can go on runs a three- or four-dog team can make, because I do half the work. Which I prefer. If I'm going to go outside in a snowstorm to be pulled around by wolves on ropes, I should at least burn some calories.

I opened the garage door and stood behind the sled. I saw nothing but darkness and the snow falling around the entrance. Already the tracks my car made were disappearing. The dogs barked and lunged in place, dying to run out into the night. I gave them the go-ahead.

"HIKE! HIKE!" I yelled and they took off. In the wrong direction. They headed for the road to the highway instead of the service road to the field trails behind the farm. There was no changing their minds and I had to stop them and manually coax them around. After some convincing, they started to head into the wilderness. With the smells of coyotes and elk in their noses, they picked up the pace.

When we're mushing, the dogs are harnessed with a single lead-section gang line and are kept together by a neck line. Jazz, my older dog, used to lead on a recreational team back in Tennessee. (Yes, there are dog teams in Tennessee. They run with wheeled carts most of the year.) So, when he's in harness he's serious. Head down, shoulders taut, his gait as steady as a metronome. Annie is a different story altogether. As far as outdoor sports go, she's more like a sorority girl in a brand-new Patagonia jacket. When given the call to take off, she bolts as if we were on a sprint line. Jazz can barely keep up with her. I always search around with my headlight, trying to find the deer she's chasing, but there never is one; she just loves to take off.

I worked up a sweat jogging behind them. Then we came to a road with a slight downhill. That's when it happened. They took off, using the grade for momentum. For half a mile or more, I was just along for the ride. At first I held on for dear life. When I'd gotten used to the speed, I relaxed, stood back on the runners, and closed my eyes. Time and space seemed to disappear. The whole world was just the snow that passed us in the glow of the lantern. Everything around us was dark and still. There was no noise in the world but the padding of the dogs' feet on whiteness and the slide of the runners on ice. No one cheered. No one barked.

I opened my eyes to see their heads bobbing in the yellow light as they loped ahead of me. The road forked and I yelled out a "Gee!" and Jazz, ever the professional, turned himself and Annie to the right. "Home for treats!" I yelled, and they picked up speed as they headed toward the garage. They slowed to a trot and I started kicking again, all of us panting and exhausted. We trudged back into the light of the garage and went through the business of stepping out of harnesses and shaking the melting snow out of our coats.

Together, the three of us made our way into the kitchen. The dogs lapped water from their big blue bowl and I put on a pot of coffee. The rest of the evening was devoted to hot beverages and library-borrowed documentaries — perfect. Annie fell asleep at my feet, and Jazz joined me on the couch with his head on my lap. I scratched his ears while listening to Shelby Foote talk about Harpers Ferry, and tried to understand what anyone does with a shih tzu.

START PACKIN'

Working house pets aren't exactly common, but I'm not sure why. My dogs have helped carry groceries home from the co-op in saddlebags, run together as a two-dog sled team, and can pull a cart of eggs and vegetables at the farmers' market. But they're also pampered house dogs. They drink filtered water. They sprawl on the couch with me when I'm reading. And sadly, they probably eat healthier meals than many children in my neighborhood do. Regardless of their sweet life indoors, though, they are a force to be reckoned with when in harness.

HISTORICALLY, DOGS WERE more than herders and hunting companions. Many large dogs were used on farms to haul carts to market, pull sleds, and carry supplies where horses and mules dared not go. Today dogs are seen less as a labor force and more as part of the family. My dogs are a living testament to the balance of pet and worker. They are my travel companions and best friends. Their light work is a huge help to me and gets them the extra exercise an indoor dog needs. If you have a dog in good health above forty pounds, you've got yourself a farmhand.

I was extremely lucky to adopt a trained working dog. Jazz was used to being in harness; he naturally pulls and loves wearing a pack. Annie, on the other hand, was a complete house pet before I adopted her, but with a little training became a great pack dog and an even faster sled dog than Jazz. The amount of fun we have out on the trails or working around the house together is uniquely rewarding because it gives me a sense of practicality and usefulness and them a chance to spend more time getting attention and being with their favorite person.

The work I'm talking about here is light and athletic. I was telling a friend about building a small dogcart and he instantly imagined a wooden box being dragged around by a rope tied to some poor beast's neck. No. This type of dog work is all high-tail, fast-trotting, tongue-out-happy stuff. At no point should your pet be uncomfortable or scared. Every step of the process should be gradual and fun for both you and dog.

WORKING DOG COMMANDS

Hike. Take off. This is the general word used by many mushers in place of the famed cry of their namesake. Actually, the old-time dog drivers never said "Mush." They used to say "March!" but the poor audio equipment of those first recordings garbled the word and the false term stuck.

Gee. Turn right.

Haw. Turn left. Why not just say "right" and "left"? Because in a blinding snowstorm with wind so loud you can barely hear yourself scream, those words sound too much alike for the poor lead dog trying to get home. The words *gee* and *haw,* however, contain drastically different vowel sounds and can be distinguished even in a blizzard.

TRAINING YOUR DOG FOR PACK WORK

If there's one thing almost any dog can do, it's help carry things in a doggie backpack. These aren't backpacks like the ones you and I know, but saddlebags designed for dogs. Whether you live in an urban community or along a dusty old dirt road, slap a pack on his back and your dog can carry water, tools, food, or a sweater and your keys. Training a dog for basic pack work isn't hard. Actually, it's a lot of fun for you and your pet, but you need to introduce it in steps. Before you take your dog to the nearest outfitter and get him fitted for a canine backpack, I strongly suggest you warm him up to the idea with some basic exercises, especially if the dog is more than three years old and accustomed to the couch life.

Start with the towel. To begin, get a standard bath towel you don't mind lending to the dog, as well as some soft ribbon or webbing — roughly two yards for a large dog. Fold the towel in half lengthwise, then again by the width, so you have a fluffy rectangle that can be placed across your dog's back, starting right above his shoulder blades. The towel should cover the upper back and drape over his rib cage, since this is how his backpack will hang. Use the ribbon or webbing and secure the towel in place. To do this most easily, encircle his whole torso with the ribbon behind his armpits and tie it above his shoulders like a bow on a Christmas package. Then bring it down past his ribs and tie it on the softer belly section. Last, run a shorter piece of ribbon across his chest, fasten it to the webbing behind his front legs, and gently secure it in place. There you have it, a training pack.

By this point your dog is surely looking at you with a mix of confusion and annoyance, and that's okay because all you're going to do with your dog is play. Take him outside for a fun walk with lots of treats and petting. Play in the yard with a favorite toy. The point is to make sure the dog is having too much fun to even notice the towel on his back. You might get some strange looks from your neighbors; just tell them he's cold or that he's having a bad hair day. You can also point at the sky without blinking and say, "There in the stars is where the Master will return," and then they'll certainly stop talking to you. It's never come to this for me, but I look forward to whipping that one out if someone is convinced I'm abusing my dog because he's wearing a pink ribbon and a towel.

Most dogs forget about the straps and go about being themselves in moments, tails wagging. But some will scratch and pick and try at all costs to bite off their burden. Don't give up right away on these guys. Most dogs that dislike the towel at first soon recognize being strapped up as a precursor to a long walk or playing with you. Before you know it, they'll be jumping at the front door when the towel comes out. Take ten minutes or so every day for towel time. Within a week or two, your dog will learn to associate the weight and straps on his back with the chance to get outside, get some exercise, and have fun.

Selecting a dog pack. After a couple of weeks of towel training, it'll be time to get your dog fitted for a pack. Some people can look at a pack and make their own out of nylon and webbing, but I don't recommend this at all. High-quality packs are ergonomically designed for a dog's comfort and load. Dogs need packs that act like extensions of their own rib cage and rest high on their back, not flopping sacks that rub their elbows or put weight on their spine.

Find a wilderness outfitter in your area that stocks dog packs (many do, now that large outfitters like Kelty and Ruff Wear make them) and ask if you can bring in your dog to get fitted. Most will welcome you and your dog with open arms if you're coming in with an open wallet. (If you don't have a willing outfitter or pack supplier near you, turn to page 185, for online suppliers that carry packs for any size dog.)

Get a pack that's labeled for your dog's weight category. Your dog might be a little standoffish about this fancy new towel, but if you've been working with him for a couple of weeks, he shouldn't mind the more efficient straps and snaps of a pack (actually, he might appreciate them). A few dog-pack features to look for are adjustable chest and belly straps and contoured sides that do not touch your dog's elbows. When you find a good fit, start walking your dog around the store in the empty pack. If it's staying in place and isn't swishing back and forth with his gait, it'll probably do just fine.

Pack training. When you have your new pack, continue with the play training. Every other walk or so, pull out the empty dog pack and let your dog associate it with fun times. Slowly and gradually add bulk and weight to the pack. A good first load is a rolled-up T-shirt in each side. Add another shirt every second or third trip. When you've gotten to the point where the pack is filled to capacity with T-shirts, you'll need to get him used to his much wider girth. Start walking him around trees and near fences or the sides of buildings; within a few bumps he'll realize without flinching that he's a couple of inches bigger when the pack's on. When your dog is completely accustomed to a full light pack, you can add weight gradually. Start with water bottles and books; as his endurance and comfort grow, add a little more. The maximum weight I would recommend is a quarter of your dog's body weight, even if he happily packs more. It's a slow process, I know, but you're guaranteed to have a happier dog if you go this route.

Please note: If your dog isn't enjoying work, don't work your dog. Some dogs, even after all the coaxing in the world, hate having anything tied to them. If yours hates having a towel tied to his back, he's not going to want to carry your carrots home from the market, either. Deal with it and buy some Milk-Bones to win back his love if necessary.

Carting with your dog. The first carting experience I ever had with a dog was with my golden retriever mix, Murray. Murray pulled a little red cart my father and I made, with thick dowel rods bolted to it for shafts. He was a big guy (about a hundred pounds), and didn't hesitate to help out with my paper route. I fashioned a candle lantern for the top of the cart, and in the dark of early morning, we'd load up with issues of the *Morning Call.* We'd walk the route together side by side, and when the cart was empty, I could ride in it like a little surrey while he trotted home. It was fun for both of us, and it made my chores something I looked forward to instead of lying in bed wishing I had never signed up to be a delivery girl.

Draft work takes a little more effort and expense than packing does. It also takes patient training. The worst thing you could do is prematurely harness up your dog to a Radio Flyer in the backyard and call her to you. She'll start walking toward you, then realize that she can't get away from the scary clanking monster behind her, get scared, and possibly run off in a panic or get hurt from the strain. After all, if you had some large, obnoxious thing tied to you for no reason, you'd freak out, too.

Fitting harnesses. To start, you need to have a properly fitted carting harness. These can be supplied only through specialty outfitters that focus on working dogs. Carting harnesses are very different from the usual pet-shop harnesses or cheap pulling harnesses. Most modern ones resemble a thickly padded racing harness with special loops and sections for traces and the cart's shaft. They are made to focus the dog's strength in the most comfortable way possible. On level ground with a properly fitted harness, a healthy Labrador retriever can pull 175 pounds, no sweat. But take that same dog and put her in a nylon pet-shop harness with the same load, and you're a few steps short of animal abuse. I cannot stress it enough — use only a real carting harness!

Once you have the carting harness (also called a draft harness), put it on your dog to take her for walks or out to play (just like we did with the training towel in the packing section). You'll want to keep things light for a few weeks. Hang it where the leashes are kept and let her know that every time she wears it, you'll be having fun together. When enough time and trips to the park have made this association a reality, you're ready to move to the next step. Most likely your carting harness came with a set of straps, called traces. These are what attach the harness to the load. Take those traces and snap them into place on the sides of the harness. Let them drag on the ground. Your dog probably will be slightly confused but not uncomfortable with them. When a few days of trace training have her oblivious to their existence, the pulling training can begin.

Pulling training. Find something fairly light and draggable to attach to the traces. I suggest a ten-pound potato sack filled halfway with dirt or sand. With your dog on leash beside you, tie the small load to her traces so that they don't hinder her natural walk. When everything is safely attached, keep the leash tight in your hand and take your first steps forward together, saying the word *pull.* Say "pull" to your dog the way you'd answer the question "What does P-U-L-L spell?" Speak plainly and directly. Take a step forward and praise your dog every step of the way with you. Walk about ten paces and then tug slightly on the leash and say "Halt" when you stop. Wait until your dog stops and repeat the whole sequence over and over. She'll probably look behind her, or she might even stop, but tell her all is well and keep walking calmly on lead.

Stop the training only if the potato sack puts your dog under serious stress. If she balks and cringes and hates your guts, you've got yourself a great pack dog. And if she hates the pack, too, you've got an even better jogging partner. Oh well, you can't teach all dogs new tricks. However, if in a few sessions your dog walks calmly with the load and gets to the point where she steps forward on "Pull" without your needing to pull the leash, you've got yourself a dog that's ready for her first cart!

Finding a cart. Dog carts, like harnesses, are a specialty item. Even if you plan on using a little red wagon or an antique goat cart, you need to have canine shafts fitted to it. Most carting suppliers have kits for these exact circumstances. If you're serious about having a working draft dog, I recommend spending the money on a professionally made cart. These are sized to a dog's height and weight, come with specially bent shafts, and work perfectly with the special harness.

When the cart is ready to be introduced to the dog, make sure it's empty and on level, paved ground. Harness her in, and praise her the whole time you're doing so. Give extra-special treats like bacon or liver, so she knows this is a wonderful event, but do not have the dog move forward yet (you just want her to be okay with being harnessed up). Give her more treats and then unhook her. That's it! Her first lesson in carting was just learning to be connected to it without stress. Go off and play now. That's plenty for today.

The next time your dog is in full harness on the cart, you can start drafting. Just like we did with that dragging sack of sand, stand at her side, lead in hand, and say "Pull." One of two things will happen. Either she'll walk calmly like she did before and barely notice the easily pulled cart or she'll panic. If she panics, stop and try again — do not praise or show any apologetic petting. That will only make *you* feel better — the dog is going to think she's being rewarded for being afraid. Try again, but slower this time. When she walks calmly, shower her with praise and treats.

Keep up the pulling and halting practice until your dog is comfortable with the cart. At that point you can slowly introduce weight, over days and weeks. When she's used to the weight and pulling happily, you've got yourself a carting dog!

TRAINING WITH THE PROS

By making a few phone calls to obedience and kennel clubs in your area, you're certain to find someone who knows someone who carts. Make a date with an experienced carter to talk over lunch or coffee. An hour with someone who knows dog sports is better than a dozen hours of reading other people's accounts. And an hour of training your dog with an experienced carting trainer is worth thousands of hours spent with books and videos.

ANGORA RABBITS: PORTABLE LIVESTOCK

When you don't have space for a flock of sheep

I WANT SHEEP. I really want sheep. Some people want to fly planes, some want to cure diabetes, and some want to run half marathons. These are all admirable goals and I wish I shared them, but all I want from my future is to sit on a hill and look at sheep. I'll know I've made it in this world when I can perch on a grassy overlook, pull out my fiddle, and play a slow Scottish reel for the flock. What can I say? When something's right, it's right, and this girl wants a lamb in her arms and a border collie padding alongside her.

Besides, you can't get any closer to the source of your clothing than hand knitting with natural fibers. I daydream about going out on a cold winter night in a homemade and hand-dyed cable-knit sweater and feeding the ewe that wore it last season. (Sheep are notorious trend starters.) The idea that a lamb can be born in my own backyard, grow up under my watch, and be shorn, and that I can take that wool and card it, spin it, and knit it into a sweater, blows my mind. The same goes for alpacas, goats, and Angora rabbits. How many other animals can give so much to us without giving their lives? I even met a woman at the Bonner county fair who raised yaks and spun vests and hats from their wool. Now every time I see a yak on TV I think, "I bet his head's really warm . . . "

UNFORTUNATELY, MY LANDLORD wouldn't let me keep sheep (never mind yaks) in the backyard. I still wanted to be a part of the whole process, from sheep to hat, but it wasn't until I met a local alpaca farmer who also raised Angora rabbits that the pieces came together for me. I thought, Now, *there's* a fiber animal that just about anybody could house and feed. Angoras produce soft, spinnable fiber and are super portable.

Rabbits also make great pets — before I got my pair of Angoras, I had a little gray lop-eared bunny named Fumiko. She was a gray mini-lop I got from a local breeder and was incredibly calm and sweet. I named her after my friend Kayo's grandmother, who lived in Japan. For the most part, rabbits are mellow, docile creatures like Fumiko. When kept clean and handled often, they're a lot like cats — they're friendly, fluffy housemates and they regard you with the same disdain (but look cuter doing it). And just like cats, rabbits can be trained to use a litter box and can therefore have free run of the house (though they've been known to gnaw on furniture and just about anything else they can get their buck teeth on). I wasn't looking for a new house pet, though, mostly because Jazz and Annie were always looking for a warm meal (remember the chickens). I suppose if I didn't live with a couple of wolves, I'd consider keeping free-range rabbits as pets. But no, I wanted fiber animals.

SINCE BARTERING is far from dead in the world of small farmers, I was able to make a deal with that alpaca farmer. I agreed to design a logo for her farm in exchange for two white Angora does (female rabbits are called does, males are called bucks). A few weeks later, we made the trade in the parking lot of a tiny Mexican restaurant in a small town about midway between our two homes. The farmer gave me basic instructions on what to feed the rabbits, how to shear them, and how to deworm them. (Sounds gross, but it's really just a quick injection.) She handed me a little bunny and showed me how to lift the skin at the back of her neck and quickly administer the medicine. The rabbit didn't even flinch. I get pretty nervous when I give blood, so I figure bunnies must be tougher than we give them credit for.

After my vet tech lesson, we shook hands and parted ways, I with two fiber animals in the backseat. Sure, they weren't the merino sheep I'd always dreamed of having, but they were livestock all the same. I smiled the whole way home. The bunnies, just a few weeks old and already giant puffballs of precious Angora fur, quietly contained their happiness (if they had any).

My rabbits lived in a hutch protected from the wind, sun, and rain on the covered porch just outside the back door. They also came into the farmhouse every now and then, to stretch their little legs and run around when the dogs were napping in another room. I set up a small pen for them in the yard, so that they could hop around and nibble on fresh grass. This combination of indoor/outdoor time gave them the exercise they needed and let them live a little more naturally than they would have been able to if they were inside a wire cage all the time. On cool summer nights, I'd start a campfire and just sit out there making s'mores with a rabbit snuggled in my lap.

Every few months, when their coats got to be spinning length (about three to four inches long), I'd shear the rabbits. In between full-body shearings, I'd groom them with a cat slicker brush, both to collect excess hair and to keep the fur from matting. I planned to collect enough wool to have a few skeins spun, bag it up, and mail it off to be processed into super-soft natural yarn for small-project knitting. Our local knitting stores carried local Angora yarn, and I used to hold it in my hands the way future architects hold blueprints, excited for what lay ahead but envious that I didn't have it for my own yet. The best way I can describe the feeling of rabbit yarn is that it's like holding wild shredded silk in your palms. Soft and hairy, but as strong as carpet wool. I couldn't wait to be holding yarn made from my own rabbits.

IT WASN'T UNTIL late fall that things got hairy for the rabbits and me. Because of the colder nights and coming snow, I moved their hutch into the garage, where I store the dogsled. During one of the first snowfalls of the season, I let Jazz and Annie into the garage to harness them up for a run.

Normally the dogs never bother with the rabbits. But on that particular night, Annie was feeling playful, dropped to her elbows, and barked at the cages. Even though she never touched the cage, the rabbits scrambled. I heard one of them scream. I had never heard a rabbit scream before, and all three of us non-

Angoras perked our ears in surprise at the uncomfortable sound of it. I harnessed the dogs, and when they were safely hitched, I walked over to the rabbits to check on them. They seemed a little panicked, breathing heavily, but were quiet and calm. No harm done. I opened the garage door, turned on my lantern, yelled "Hike!" and we headed off into the night on the dogsled.

LATER THAT EVENING I went back to check on the rabbits. While changing their water bottles and scooping out feed pellets, I noticed something odd. One of them was dragging her back leg, as if it no longer worked. I couldn't tell if she was still shocked from Annie's barking or if she was really hurt. I gently lifted her up and, although she didn't wince or cry, her back legs seemed limp. It was after ten o'clock by this point, so I decided to call the vet first thing the following morning if she wasn't back on her feet.

The next morning, I walked into the garage and witnessed a horrible sight. The rabbit with the broken leg was now missing a foot. Either out of her own madness or the viciousness of the other, she had nothing left but a bloody stump with exposed bone and sinew. It wasn't until then that I put it all together. When Annie barked and the rabbits scampered, this one must have caught her paw on the wire cage floor, and in her frenzy dislocated a vertebra, paralyzing herself from the waist down.

Overcome with guilt and fear, I grabbed a clean cardboard box, filled it with fresh pine shavings, and put in bowls of food and water, then separated the two rabbits (in case cannibalism was the culprit). The doe with the broken foot could hardly move. She was still alive, but just barely. I ran into the house and called the closest veterinarian and told him what had happened. He explained calmly that the rabbit would need to be put down; it had a broken spine and was now in great pain. He told me I could bring her in for an injection, but moving her in and out of the box might cause her unnecessary pain. The best thing to do would be to put her out of her misery as quickly as possible. He was clear: I should kill my rabbit.

WITH WHAT? I didn't own a gun, and I wasn't about to slit a fluffy bunny's throat. I called Diana, waking up her family on a Saturday morning. I explained the situation and asked nervously if I could borrow one of their rifles; I knew they had a rack of .22s and shotguns by the back door of their kitchen. Her son, Zach, had used them to take out coyotes that were stealing chickens. Since she was in the business of raising and harvesting her meat animals, I thought she could do it for me. She told me to come right over; they would help me take care of this.

I carefully moved the rabbit into the back of the station wagon, covering it with blankets to protect it from the cold and making sure the box wouldn't be jostled too much. I headed south on the highway toward Floating Leaf Farm. As I passed the post office, I remembered driving down this same road in early spring, happy and singing, on my way to pick up a carton of chicks. Now I was carrying to its death an animal I had raised myself. Few car rides were as quiet and somber as this one. It started to snow.

I PULLED UP to Diana's back door, left the engine on so the car would stay heated, and let myself in. It was pretty early in the morning for an execution, but the family was up to meet me. Diana came out and gave me a hug. Her husband, Bruce, had a pump-action .22 rifle loaded and waiting for me.

We were getting ready to head outside and I needed to make sure I wasn't the one pulling the trigger.

I asked sheepishly, "So . . . who's going to take care of this? I mean, I've never shot an animal before. Maybe you should do it, so it goes as quickly as possible for the doe."

Diana wasn't going to let me shirk my duty as a farmer. "Well, it's your rabbit. You raised it, held it, fed it. It knows you and you're the one it trusts to take care of it."

I was scared, but Diana wanted me to understand the full weight of responsibility that keeping livestock entails — that it's not all fluffy scarves and omelets. That sometimes a rabbit or chicken would be beyond medical help and need to be put down. She wanted me to understand it isn't fair or realistic to get into farming and expect someone else to do the dirty work for you. Besides, someday something like this could happen again and she might be out of town or unable to help. She was right, and I was out of line. Diana picked up the rifle, and the two of us stepped outside into the gathering snowstorm.

Diana dug a grave in the nearly frozen earth and I gently laid the rabbit inside. I cannot tell you how sobering these simple actions were. I don't take death lightly; I won't even eat meat because of it. I've been a vegetarian for years for many reasons, but none as strong as the words the Buddha said thousands of years ago:

All beings tremble before violence.
All fear death.
All love life.

I cocked the rifle. Aimed. Fired.

The rabbit died instantly. I stood in the snow beside Diana, shaking. I had never killed anything before. I ate plenty of hamburgers and bacon before I became a vegetarian, but I never had to be the one ending the life of the animal that gave it to me. I had been the cause of so much death, and all the animals that fell for my plate seemed to be watching us in the snow at that moment. I'm not opposed to animals eating other animals. I understand a human's role in the food chain. But being the reason for another animal's death, even that of a small rabbit dying in terrible pain, was hard. Really hard.

There was an unofficial observed silence, but it didn't last long. Next to us in a stack of hay bales covered by blue tarps, a hen cried out. After months of living around chickens and hearing all kinds of clucking, I knew exactly what that specific cry meant: she had just laid an egg. There in a snowstorm, life and death were coinciding. They were just a rabbit and a chicken, I know. But they aren't just rabbits and chickens to the people who live with them, raise them, harvest them, and fall to their knees in gratitude for them. They're much more — at least, at that moment they were. The whole world snapped back into some comprehensible order and, with glassy eyes, I went inside to warm up.

Inside the Carlins' farmhouse, we talked seriously about how utterly heavy the "simple life" could be. How common death is. How vicious animals can be to each other. So many books about hobby farms and specials on PBS seem to gloss over this reality, ignore the basic responsibility of it all. You can be a vegetarian on a hobby farm and eat all the Tofurkey you want, but those animals you raise will get sick, get hurt, or die of old age. Every thing you grow attached to — animal or other — will someday meet its end.

Sometimes the only responsible thing to do is to be the person who gives it that end as humanely as possible.

I looked out the window and saw the snow was coming down harder; the weathermen were calling for a serious storm to roll in. And when they say "serious" in the Idaho panhandle, they mean well over two feet of snow in a night. We ended up getting nearly three feet by midafternoon of the following day. What if this rabbit had needed to be put down just twenty-four hours later? I wouldn't have been able to reach Diana's farm. And not even an emergency vet would have been able to get to my farmhouse if the weather was truly awful. That rabbit might have been eaten alive in another day's time. It weighed horribly on me.

On the way home, I stopped at a sporting goods store and for the first time in my life bought a gun. The small store was owned by a hunter, and besides selling brand-new guns, he also sold refurbished and used rifles from locals and estate sales. I purchased a vintage Remington .22 bolt rifle and a box of bullets. I took it home, shot it once to make sure I knew how, then emptied the rifle, put the safety on, and set it in a corner of the living room.

How simple was the simple life? Clearly, it's complex enough to make a Buddhist vegetarian kill a rabbit at point-blank range, then go buy a gun. Your lifestyle preferences are not considered when it comes to caring for the lives of others on a farm. Not everything can be as simple as we'd like.

FEED, SHEAR, SPIN. REPEAT.

Although this particular part of my small homestead's story was sad, not all was lost. I still had a healthy Angora left, and she continued to produce great-quality white fiber, part of which I was able to sell online to other spinners. Because they're so easy to raise and care for, I still recommend Angoras for wannabe shepherds.

FINDING YOUR OWN HERD

Angoras are a specialty breed, generally raised by a small number of fiber lovers. (Many Angora breeders also have sheep, alpacas, or goats for the same purpose.) To find an Angora breeder near you, start by searching online. Asking around your local ag community or looking through national breeder directories can also point you in the right direction. Another option is your local knitting or specialty yarn shop. If it has local suppliers, it may have all the contacts you'll need to get in touch with a nearby rabbitry.

CARING FOR RABBITS

Keeping Angoras requires little previous animal-care experience, but still calls for research, time, and energy, as well as the tolerance for a scratch or two (rabbit claws are sharp, even after being trimmed). They're a great starter animal for people who think they might want goats or sheep someday but want to start small. If you raise Angoras, you'll learn all about health, handling, shearing, care, and feeding, and come county fair time you can try to earn a ribbon or two, if you like.

Housing. If you plan to house your rabbits outdoors, you'll need something sturdier than those cages the pet-shop people sell. A wire hutch with a solid roof and sides will protect your rabbits from the elements and from predators. You can buy one that's preassembled or you can build one from plans or a kit. To give the rabbits a place to feel safe and protected, get a hutch that has a "den" (a separate, enclosed area where they can retreat, especially when they're raising a litter of kits). You might be tempted to give your Angoras some straw for bedding, but it's actually not a good idea. Hutches are designed to let droppings fall through the wire floors to keep the animals clean (you can compost the waste and use it to improve the soil in your gardens). Also, the bedding material will make the wool become matted, making it hard to spin.

Food and water. Make sure you offer your rabbits a high-quality specialty feed made just for rabbits. Rabbit-food pellets come with the proper balance of proteins, grains, and ground vegetables, and are formulated to match a natural diet they would choose themselves in the wild. Lower-quality feeds with less protein won't meet a fiber rabbit's dietary needs — 16 to 17 percent of its meal should be protein, and that's especially important for animals that are being raised for fiber. Good food equals good wool.

Rabbits are creatures of strict habit and prefer to be fed at the same time every day, regardless of whether it's at night before bed or before your morning coffee. The occasional treat of fresh vegetables is always welcome.

Grooming. Grooming your Angoras might seem a little intimidating when you get up close and personal with your own ten-pound balls of fluff, but truth is, it's pretty low key. They require only basic brushing and nail trimming (use a cat claw trimmer). You should plan on brushing your rabbits with a small slicker brush once a week, and bagging any loose hair in a plastic bag or sealed container. This is the only way some people collect fiber. I can't blame them — the full-body shearing process is tricky. But since I plan on shearing my own sheep someday, I figured I might as well learn on a rabbit.

Shearing. When you decide to harvest the fiber from your rabbits, you need to decide how. The choices are to collect it through combing and brushing or by shearing or plucking your animals during their natural molting cycles. You can have at it with scissors and shear them the way you would sheep. Plucking produces the longest and best hair. When it comes time to shear (when their hair is at least three to four inches long from base to tip), make sure you're cutting them only with hand shears (sharp scissors) and not an electric grooming tool like the ones people use on miniature schnauzers. The first time you shear a rabbit, I strongly suggest you have the guidance of an experienced Angora breeder, but for those of you not so lucky to have that luxury, here's a basic walk-through.

Set the animal on its back in your lap (rabbits are very submissive and will let you do this with little resistance) and start by gently cutting the hair in small sections around the belly and teats.

Stay close to the skin, but never touch it (the last thing you want to do is cut the rabbit's skin). If you need to do it in shifts to stay focused, shear the rabbit over the course of an entire week: do the belly one day, the right side the next, and so on. You'll harvest the hair off the entire animal except for the neck, feet (the fur acts like booties and protects their feet from the wire cage floor), and tufts on the head and ears. A shaved rabbit looks hilarious, but if you do it right, you'll have a few small bags of the softest natural fiber anyone could buy.

SPINNING FLEECE

I learned to spin wool from fiber when I was seventeen years old, at my hometown's annual heritage festival. A woman had a big wheel between her quilt-skirted legs and was taking those long, thick spools of roving (carded wool) and pedaling them into chunky yarn. She must have noticed my unusually high level of interest, and she asked if I had ever spun wool. I just shook my head and kept watching. Then she asked if I wanted to learn. Of course I did.

She handed me what looked like a poorly designed spinning top: a nine-inch-long dowel with a hook on the end of it, capped with a six-inch wooden disk. She said it was called a drop spindle, and she showed me how to spin it and slowly feed it the roving and turn it into yarn. She gave me a homemade version of her drop spindle — a craft-store dowel rod and an old AOL sample CD (so that's what people had been doing with those) — and a chunk of raw fiber and told me to buy two dog slicker brushes and card it with those. I left her little booth a changed girl and spun my first scrappy ball of yarn that same day.

Drop spinning Angora fleece, however, is best done after it's been blended with another fiber, such as sheep's wool. Angora, though very soft, is also very inelastic, which makes it difficult to spin. In order for me to use the fleece from my rabbits, I'd need to send it off to be processed into yarn or, at the very least, blended with wool and turned into roving.

RAISING RABBITS FOR SALE

Of all the livestock animals, rabbits may offer the widest variety of uses. They're not just fiber producers — they're also raised for meat, for pelts, and as pets. Some folks raise large breeds like New Zealand whites to sell to gourmet restaurants. Rabbit is a specialty dish that's gaining more and more popularity every year. Even a person with limited land and Research, Son could raise a hundred pounds of healthy, organic rabbits for his or her local chefs. Or, if slaughtering rabbits doesn't appeal to you (something I can empathize with), you could raise kits (baby rabbits) to sell as pets or Angoras for fleece.

HOMEMADE MOUNTAIN MUSIC

Homegrown entertainment, unplugged

IN TOWNSEND, TENNESSEE, there's a dulcimer shop in a little cottage just down the road from Smoky Mountains National Park. The man who owns it builds his instruments on-site and runs the place with the help of his family. He's everything you'd expect the proprietor of a mountain music shop to be — soft-spoken, unassuming, and humble. Every weekend of the long southern summer, he puts on free outdoor concerts behind the store. There's a small wooden stage called the pickin' porch, which enjoys the cool shade of fat leafy trees and the accompaniment of a bubbling creek. Local musicians come to perform traditional Appalachian tunes, and any polite person with a blanket or lawn chair is welcome to listen in.

My first summer in the South saw a lot of these performances; they became staples of my Saturday ritual. After a long week under the awful fluorescent lights of the office, I'd pack up the dogs and a sack lunch and head east from the city into the park. I'd spend the whole day exploring and hiking, and come dinnertime I'd be exhausted. But I always knew that on the drive home I could pull into the small parking lot of the dulcimer shop for some of that sweet mountain music, and I was grateful just to sit and listen.

Because it was my last stop before I had to drive back into the city, I savored it even more. I'd buy a cup of coffee and a fried apple pie from the neighboring church (which set up a food stand during the concerts), find a friendly patch of grass or a tree to lean against, and take in the show. As the musicians played the old ballads and hymns of the South, I sipped my coffee and watched the fireflies in reverence. The combination of sweet dulcimer music and a few miles in your hiking boots is more potent than a horse tranquilizer. Instant calm washed over me. The nearby creek was just loud enough to play background percussion to the strings. Let's just say that if you're in search of heaven on Earth, I know whose backyard you should be in.

SHORTLY AFTER THOSE first few concerts behind the shop, I bought a mountain dulcimer of my own. It was a student model, made over the mountains in North Carolina. It wasn't fancy, but it sounded wonderful to me. After a little research, I discovered that Knoxville had a dulcimer club that taught new players on Sundays in a Methodist church hall for the flat rate of fifteen dollars a year (by god, that's a steal). I learned to play from their patient volunteers and with the help of books, recordings, and other self-instruction aids. In a few weeks, I had learned a couple of songs and could play them with my eyes closed. If having a dog helps make you feel like part of the Knoxville community, learning the dulcimer makes you feel like part of its history. The dulcimer changed things for me. I didn't feel like a tourist anymore. I felt like a real highland girl who could strum out a fast-paced version of "Old Joe Clark" with the best of them. It gave me a sense of place and local pride in a region where, just a few years before, I never knew I'd live. That's a lot to get from a funny-shaped wooden box on your lap, but there it is. And the best part about learning to play is that wherever I go, whatever happens, I know that music is with me. The smiling face of that shopkeeper comes back every time I play my dulcimer, and I play it partly for him. Some things you really can take with you.

Right before my big move out west, I took my dulcimer back to the shop for one last going over. The white-bearded proprietor stood behind the counter, sporting his signature blue-jean overalls. I put my instrument on the countertop and asked if he would please put on some new strings and give it a once-over. He said that it looked just fine and probably didn't need any special care, and that he wouldn't feel right charging me to do something unnecessary. His livelihood was to tend to these very instruments, and he was passing up the chance to have someone pay him to tend to one. It broke my heart. I was leaving a place where dishonest money wasn't wanted.

Then I explained that I was moving and wouldn't be back for a long while, and wanted to be sure it was as good as new before I skipped town. He saw in my face that I just needed his blessing, and he took it in the back to tinker with. While he was gone, I wandered around the shop, admiring the fiddles and mandolins on the walls, like they were installations in a museum. A few moments later he had it ready for me to zip back into its case.

While I was settling the ridiculously low service bill, he asked where I was moving. I told him I was going to Idaho in a few days, that I'd gotten a job out there. There was an uncomfortable silence that followed this declaration. He looked at me like I'd just told him my puppy had drowned. He put his hands on the counter, leaned over with the presence of Gentle Ben, and said in a tired but utterly serious tone, "Now why would you want to go and do a thing like that?"

He followed this with a smile. I smiled back, but his meaning was understood. Dulcimer players who found their sound deep in these mountains weren't supposed to just up and leave. That kind of adventuring wasn't necessary for someone with the sweet music in her lap (in 1849 or presently), and it just plain annoyed him. I liked that about him.

The day I left for Idaho, my dulcimer was one of the few things I packed in the station wagon. It rode out of Knoxville with the dogs and me; it had become part of our family. I brought it into the hotel room with us every night. I didn't want to feel alone so far away from home.

A FEW WEEKS LATER I was well into my new adventure at the very tip of northern Idaho. The weather had been in the fifties when I left Tennessee, and now the average temperature was somewhere down around go-outside-and-die-alone-by-a-pine-tree. All I saw outside the farmhouse windows was a blanket of ice and snow. Beautiful, but extremely confining for the new kid. Since I didn't have a ski pass or any friends yet, I was pretty much spending my free time by myself, indoors at the farm. So, I watched movies, read, plucked the dulcimer. But that wasn't enough to temper the anxiety of being a stranger. Maybe it was boredom or maybe homesickness, but whatever the motivation, I went online and ordered a cheap fiddle. After all, I promised myself when I was in Tennessee that I'd learn to play the dulcimer, so why not make the same deal with the fiddle in Idaho?

I had always wanted to learn feral violin but thought it was something that took years of apprenticeship, so I never really tried. I received one as a Christmas present back in high school, but I think I gave up on it before I even gave it a real go. I mean, come on, weren't fiddlers people raised at the knees of their pipe-smoking elders along the Blue Ridge? This wasn't something I could just do, was it? The dulcimer was one thing, but if it was a fat happy pony in a pasture, the fiddle was a Thoroughbred chasing foxes in the pines. This was no idle pursuit for a bored girl in Idaho.

Too late. There was no helping it. I wanted to learn. I *needed* to learn. I looked up to fiddlers, as if their ability to whip out a jig was nothing short of rustic enlightenment. I loved the emotion that could be drawn out of a fiddle, like the sound of a person laughing or crying. So, I didn't care anymore how hard it would be to learn. I had a cheap fiddle coming in the mail, and I'd teach myself. I ordered a book and CD from a man who taught fiddle down in Asheville, North Carolina; he promised to teach anyone old-time fiddling, regardless of their experience level.

My fiddle came on an icy day in early February. Pulling into the driveway, I noticed a box shaped like a miniature casket propped up against the farmhouse door. I had ordered the sorriest student fiddle ever made. The instrument, case, and bow only cost twenty bucks, but by the way I carried it into the house, you'd think it was an ancient sarcophagus. This wasn't an instrument, it was a promise. I was going to learn to play it, and the music that came out of it would take me back to those firefly groves of Tennessee while I learned to love my new home.

To summarize a very tedious and squeaky section of this story, I learned to play the fiddle. It was slow going at first, but all through the winter, spring, and summer I played that cheap fiddle every day. To its credit, it rarely lost its tuning and never broke a string. The book I was learning from was golden: *Old Time Fiddle for the Complete Ignoramus,* by Wayne Erbsen. It taught fiddling by ear, so I never had to fuss with actual notation. Mr. Erbsen was funny and honest about the process, and within a few weeks of playing my first scale, I had memorized a handful of songs.

I wasn't a prodigy but I was playing, and that's all it took to keep me going. I played at friends' bonfires, or outside by my lonesome on the rundown blue Ford tractor while the chickens pecked around the yard. I learned about thirty old-time songs, and after a few months of religious practice I got to the point where I could listen to fiddles in movies or on recordings and copy what they played. I started playing along with favorite songs on CDs, adding the fiddle where no one asked for it to show up. I'd sit furrow-browed by the computer, playing pop CDs, trying to figure out in my head where a fiddle could be worked in. It was like a math problem, but entertaining and applicable. I still loved my dulcimer, but this was unbridled passion. I was addicted. I was in love. I felt like a fiddler at last. And then the real test came.

. . .

Every fall and spring in the foothills of the Smokies, the Old Timers' Festival takes over the national park. It turns an otherwise borderline-campy visitors center into a national treasure of historic song and dance. The air itself is heavy with dorian undertones and clawhammer banjo flailing. People come from all over to jam out bluegrass and mountain tunes, catch up with friends, and eat great food. Strangers become bandmates and play the songs that their Scotch-Irish ancestors handed down through the generations, songs like "Barbara Allen" and "Pretty Saro." This was the very event I was flying back east for. After nearly a year away, I was returning with a fiddle on my back and a solid foundation of mountain music in my head. Walking around the festival grounds felt like I'd been away on vacation and was finally coming home.

I carried my fiddle in a makeshift backpack I'd sewn together from an old messenger bag and a retired daypack. I had my bow at the ready, like a sword, in my right hand. I wanted to be a part of the strum-and-pluck scene. I'm not talking center stage — I'd have been happy to find a corner by myself and just be a lone fiddler playing "Wayfaring Stranger" to a few passersby. But, it turns out, if you're in Tennessee and you have a fiddle on you, people want to hear you play. I played at picnic tables and in line for blooming onions. I played waiting for the bathroom and in gift shops. When my friends and I walked into the visitors center, an older woman asked me if I would play for the people manning the arts-and-crafts booths. I started playing and was keeping up with myself just fine when an older lady slowly approached me with her walker. As I kept playing she whispered in my ear behind a raised hand, "Cluck Old Hen," which was the name of the song I had been sawing out. She knew it and wanted me to know she did. As if it was our little secret, a young Yankee and a Tennessee old-timer whispering hundred-year-old song titles in the shadows of the great rolling Appalachians. I don't think I ever smiled as much in my life as I did right then. That was when I knew I'd never stop playing.

My relationship with old-time music happened by chance and proximity. Had I never moved to Appalachia, I don't know if I would ever have had the drive to make mountain music my own. Which is pretty amazing to me, because there isn't a country skill I've learned that I care about half as much as those old songs. For me, learning the music of the original homesteaders is just as important as learning how they lived off the land. The musical heritage of the pioneers is as authentic and accessible today as their recipes and quilts are. Just like their domestic arts, their music continues to transcend time and experience. It's been hundreds of years since the cabins of the pioneers first sent smoke out their stone chimneys, and even today you'll find thousands of artisans who keep the elements of their culture alive — from cast-iron stoves to cookbooks and natural-fiber dyes, the old ways are alive and well. The same goes a hundred times over for their ballads.

START PICKIN'

Making music is the ultimate in self-reliance. As soon as you hold that banjo in your hands, you've opened yourself to a tool that can be all the therapy, companionship, satisfaction, and dedication this world has to offer outside of dogs. The rest of this chapter will show you how to get started: how to choose an instrument, teach yourself, and start a local jam. May that wonderful music that once wafted from campfires everywhere now float up from our fire escapes. May the fiddle rise again.

SOME WOULD argue that playing an acoustic instrument isn't a practical skill. Well, they can argue all they want — I can't hear them over the fiddling anyway. I'm stubborn about adding music, particularly old-time music, to your life, because it has the ability to fix things. Things like boredom, anxiety, low self-esteem, and loneliness. This reality is understood by people who have been playing their whole lives, but to those of us who arrive a little late to the game, all those benefits come as a shock. There have been times when I was downright miserable, but when I came home at the end of the day (and there's always an end, even to the worst of days) and walked into my living room to find my banjo leaning against the couch, I knew my dark mood was passing. Steve

Martin was right when he said it's impossible to be miserable and play the banjo at the same time. Those old tunes can revive you even if you're deep in despair and want nothing to do with positivism — they'll hold your hand and sound just as mournful as you feel. It's free, productive therapy.

Creating your own music is kind of like growing your own food. You can grow simple things or complicated things. You can put as much effort into your garden as you can muster, but at the end of the day even a single sunflower in a pot is still nice. When you're playing an instrument, what matters is that you can make a song without the use of microchips or outlets. The ultimate musical self-reliance needs nothing but yourself and your instrument to entertain anyone within earshot.

CHOOSING AN INSTRUMENT

The first step in playing mountain music is choosing your partner in crime. The tools of the trade are the fiddle, banjo, guitar, mandolin, and mountain dulcimer. These instruments have four things in common — they are stringed, cordless, wooden, and portable. This makes them perfect for picking up anyplace you feel like lifting your spirit. You don't need to plug them into anything because they run on renewable energy (you), and if they're treated well, good-quality acoustic instruments can last a lifetime.

If you're not sure which direction to turn, here's a primer. When you know a little about each instrument's history, you'll have a better idea of what suits you and how to start researching.

The mountain dulcimer. I've got a special place in my heart for this awkward-looking but beautiful-sounding creation. What we now consider a staple of the Appalachian gospel sound is actually derived from European settlers in Pennsylvania (I can relate — so am I). Today you can't walk into a bluegrass store without seeing these hourglass-shaped instruments hanging from the wall. Besides being fairly inexpensive, they're crazy easy to learn. Dulcimers are set up chromatically, so every fret you strum is a perfect chord. This means you can play anything on it and it's in tune and makes sense. So, it's beginner-proof. If you've never played an instrument before and want near-instant results, this is the instrument for you.

The fiddle. The fiddle is the avatar of old-time music. Because the fiddle is small and travels well, it was one of the few possessions brought overseas by the original European pioneers. Old ballads, reels, jigs, waltzes, and hornpipes were the bread and butter of fiddlers back then, and most other instruments were playing backup to it because of its sheer popularity. Those original players had more than a violin; they had a time capsule and passport all in one. The songs took them back in time to the old country and helped keep their traditions alive in a strange land. The fiddle could start up a dance on the spot, and for that reason some strict members of the church called it "the devil's box." When it partnered with a banjo, people really got rowdy, so together they were known as "the devil's stalking horses." Well, damnation aside, it sounds as good today.

The violin and the fiddle are the same instrument. The only real difference is the type of music that is played on them. Violins create those haunting classical and contemporary pieces, and fiddles pretty much just went to hell in a handbasket. So if you've wanted to play the fiddle but only have your granduncle's violin in the attic, you're in luck. Fiddles have four strings and are played with a bow any way that's comfortable for the player. There's nothing fancy to them, and fiddlers disregard most of the etiquette and rules violinists need to deal with.

The banjo. Some people say the banjo (originally called the banjar) is the only truly American instrument. While the instrument we hold in our hands today is an American creation, it's actually a modern version of an African instrument that plantation slaves re-created in America. They fashioned this strummable drum on a stick. When those banjar tunes butted heads with the sounds of Irish fiddles and Italian mandolins, a new mountain sound was formed. The mixing of strong African rhythms with the fast-paced melodies of traditional European ballads created a distinctive sound that is the cornerstone of bluegrass music heard today — proof positive of what amazing things can happen when cultures collide.

The standard banjo has five strings: four along its neck and a shorter one that's attached closer to the base. Old-time music is usually played with this type of banjo, but there are all sorts of variations of the animal. Four strings, tenor, resonated, mandolin styles, the list goes on. Banjo players are great partners in crime for fun, lighthearted folks. The signature twang brightens up any jam as soon as a player starts plucking. If you've played guitar, it's easy to make a smooth transition, since even basic strumming can bring a smile to most people's faces. If you want something in your hands with the meat of a guitar and a cult loyalty the people at Lexus envy, you might want to look into the banjo.

The mandolin. This is the Italian import that came to America in much the same way the Scotch-Irish fiddles did. Small and portable, it accompanied Italian immigrants making their home in the New World. This small, eight-stringed, guitarlike instrument has a high-pitched yet still mellow sound that has become synonymous with modern bluegrass. What was once a humble backup to old-time fiddles now steals the show in many a bluegrass band. Mandolins are hardy, beautiful, and small enough to be stuffed in a backpack on a long hike into the mountains. If you have nimble fingers and can hold your own in a fast-paced conversation, I'd say the mandolin might be the perfect fit for you.

The guitar. The guitar was for me, and has been for countless others, an introduction to the world of acoustic strings. My high school had four options for a mandatory music class — band, choir, theory lectures, or guitar lessons. Guess which one I chose? In my guitar class I learned to tune, hold, and strum chords and even got to teach everyone to play a simple rendition of "Leaving on a Jet Plane." Good times.

The guitar is the instrument you're probably most familiar with. It really doesn't need an introduction, since it's been a staple of modern music so long we forget that people didn't even see them in America till after the Spanish-American War. Soldiers who spent time in Cuba came home with them, and their deep resonance, sweet chords, and easy pluck-and-strum style came too. Guitars have six strings made of nylon or wire, and get their deep tone from the strings' vibrations resonating from those giant wood bodies.

When you've made your decision, guess what it's time for . . .

Research! That's right, grab a backpack and jump on your bike, because you'll be picking up recordings, books, journals, the works. You'll be taking notes, making phone calls, and reading magazines long before you tune anything (unless you're foaming at the mouth and need to make music tonight — in that case, skip ahead to the Start Today: Strumsticks section at the end of this chapter). Everything you need to get started is waiting for you right now for free at your local public library. Satiate yourself with old-time albums and recordings. When you drive to the grocery store, go out for a jog, or sit at the computer at work, listen to the recordings of the instrument of your choice. You'll create a subconscious climate of music to aspire to, and when you start churning out tunes, you'll be surprised how many of those techniques and tempos stuck. If you're a college student, you might have more opportunities than you can handle with your music department. Clubs, lessons, and other people your age just starting out are everywhere.

After you've been exposed to the music and have your instrument in hand, you're ready to start teaching yourself. This is the point where you either become a zealot or lose the faith entirely. In my experience it completely depends on your instructional materials. A bad beginner's book can crush your musical dreams faster than any critic alive ever could. If the books you use are boring or too heavy on theory, get rid of them. What you need is a friendly introduction that keeps you inspired. Companies like Native Ground Music in Asheville, North Carolina (see Research, Son, page 189), publish books just for people like us. Their Ignoramus series is solely responsible for everything I learned on the fiddle and banjo. A casual, folksy text and audio recordings every step of the way got me through two winters of learning new instruments without paying for outside lessons.

I can't stress enough what a bad idea it is to use those cheap "learn-to-play" DVDs sold in the music store. They show experts playing too well and too fast. It's ridiculously discouraging to see someone play like a pro when you don't even know how to hold the damn thing. I guess they're useful in that respect: you can see people playing correctly and holding things right. So as far as visual aids they're gangbusters, but when it comes to learning from them, it's like being the fat kid on the sprint line.

JOIN IN A LOCAL JAM

Playing with others is like a musical workout. It's the backyard pickup football game of bluegrass. When you're playing the basics with confidence, you're ready to find other people to jam with. I know that might seem superfluous to some, and I was originally of the same mind-set. I wanted to learn to play for my own amusement. But the nature of this style is communal, and the things you can learn in half an hour with other musicians can cover what might take you months of plucking alone. Old-time and bluegrass jams are happening in every county and on every college campus in America. You just have to know where to look. Ask around at the local music store, or search online for musician meet-up groups (see Research, Son, page 191). Show up with an open mind and a giant sense of humor, and you'll love it.

It might be rough going at the beginning — the first real bluegrass jam I joined was intimidating. People flat-picked through songs I had never even heard of, much less could keep up with.

But I kept trying, going back, and after a few weeks I was adding to the group instead of detracting from it. No one gave me dirty looks or asked me to shut up. They showed me where to print out free sheet music online and how to follow chord progressions, and together they got me up to speed.

START TODAY: STRUMSTICKS

If you want to start making music today but have no idea how to play anything and think that even after a cute story you won't be able to learn, have I got the instrument for you! Ladies and gentlemen, meet the Strumstick — the love child of the dulcimer and the banjo.

It's shaped like a long, skinny guitar but only has three banjo strings and is fretted like a dulcimer. What you end up with is the easiest mountain music–making machine ever created. Since dulcimers are fretted (frets are the metal lines in the neck of the instrument; you press the strings against them to get a note or chord), as long as you're holding down a fret and strumming, you're in key. And the three banjo strings sound as wild and earthy as anything else on the music-store shelves.

The really nice ending to this little story is that Strumsticks are fairly inexpensive. Expect to get a brand-new one for around a hundred dollars and used ones for half that. Most music stores carry them or can order them for you if you ask, and many outfits make them, but the real deal is the McNally Strumstick (see Research, Son, page 190), which I recommend if you want something to play tonight at the campfire.

MOVING ON

So, I landed in Vermont

ON A PARTICULARLY NASTY NIGHT in February, I found myself driving over the Fourth of July Pass in western Montana, at the start of what was supposed to be the next chapter in my life: a cross-country road trip to New England, where a new job waited for me. But so far, the miles logged back East were hard going. I was driving through what may have been the worst road conditions I'd experienced in my life. Book analogy in play: every time the car skidded near the edge of the pass, I was given a healthy reminder that this might be the *last* chapter.

The weather belied the road's happy namesake. I cursed myself for leaving Idaho so late in the evening, especially when I'd seen the swirling winter clouds of an oncoming snowstorm earlier in the day. I'd foolishly assumed I would beat the worst of it, and had left the state with some false confidence. Hard, white balls pelted the silver hood. A mocking squall of powder rocked the car. Too late to turn back now. My whole body felt hot — the kind of hot that isn't associated with stress, but with a bad case of the flu. Wonderful.

I expected to slide off the road at any minute, and getting to know what the bottom of a ravine tastes like was becoming more and more of a possibility. I drove as though I had blinders on, my eyes ahead and aware, my heart pounding. The stress level could've knocked out Secretariat. I was so focused on not killing myself, along with Jazz and Annie, that I didn't realize I wasn't breathing properly. I kept yawning. Thanks to lack of sleep and heightened anxiety (I was wound tight as baling wire), my body demanded that I gulp in air to get some oxygen flow to my brain after miles of shallow inhalations.

Ever so slowly, I rolled that dented wagon up the mountain. I silently prayed I'd just make it through the pass and be back on the highway. In my entire life, I had never hoped so badly for level ground and streetlamps. I wanted to pull over and scream. Let's all agree, I have had better nights.

The blizzard was getting worse and I had more than 2,000 miles to go before I reached my new address in Vermont. I softly, desperately cursed the snow on the other side of my windshield. It's not that I dislike snow. I adore it. It's just that I'd rather adore it from the back of a dogsled or beside a stocked fireplace — preferably with a parked car with studded snow tires hibernating in the driveway.

A Suburban passed me, its driver laughing as he flew past my Sherpa-wagon. I must've looked quite the joke — barely moving uphill, sliding around like a rodeo clown on Percocet. Anyone who had peered into my vehicle that night would have seen two dogs, blissfully ignorant of their impending deaths, sprawled among a menagerie of piled suitcases and instruments. A folded dogsled lashed to the roof (which delivered small peace of mind; if my front end decided to make out with a snowbank, at least we could mush to safety). Stickers, half peeled off from the wind, flapped along the sides of the wagon, touting places in states I no longer remembered.

To temper my anxiety, I turned on the radio to the station suggested by the (barely visible) flashing road signs. The voice on the emergency station announced that the Fourth of July Pass would be closing shortly. They just had to get some stragglers off the mountain before it was declared unsafe to pass.

As the miles racked up on the odometer, I slowly came to the crushing realization of everything I'd left behind me. (In hindsight, this may have been overly dramatized, given the circumstances.) The heartbreak of leaving friends and experiences was compounded by the reality that I had no idea what lay ahead. All I knew was that somewhere in southern Vermont, a new home was waiting for me. A new career with lots of strangers in adjunct workspaces was waiting, too. I'd heard warnings about the aloofness and coldness of New Englanders — that getting invited to dinner only happened if you made reservations at a restaurant and the hostess asked if you were ready to be seated. I was accustomed to the open warmth of the people of Tennessee and the spunky wildness of the people in Idaho. I worried I was driving into a club I had no business being a part of, and the locals would let me know it. Although I was still physically in the inland Northwest, I already missed it terribly. It's hard to see the yellow lines when you're wiping away tears. The dogs were silent as graves.

Driving alone at night does something to you, anyway. If you're like me, it welcomes that self-involved melancholy you can only truly attain by driving alone listening to slow music. I had made a mix-tape of songs as a going-away meditation. Some of the songs were happy — music I'd played around campfires and at jams in town. Others were deeply moving. Currently, Josh Ritter's *Idaho* was softly crooning. I had heard this song just a few months before, sung by Josh himself at the Festival at Sandpoint. There, on the banks of Lake Pend Oreille, I sat on a blanket with neighbors and friends, listening to that homily for the Gem State. It was one of those high summer days you remember in saturation. Grass is greener, the sunsets redder, the mountains a little higher. Above the crowds, in the stadium lights, a small flock of ospreys flew from nest to nest. (Maybe putting up 100-foot-tall lights in prime nesting territory wasn't such a genius idea. The local ospreys had moved into their new condos, and dimmed the lights with their stick nests.) Now, a half turn of the wheel later, I was driving away to that same music. It reminded me of the going-away party my friends had given me just a few nights ago in town.

My favorite place in Sandpoint was a little bar called Eichardt's, a small skiers' bar right in the center of town. It was always dark inside, but never depressing, the kind of hole-in-the-wall place that isn't so much a hole as it is the gateway to your personal Narnia. Our place to blow off steam, see live music, grab a Guinness, and eat the most amazing garlic fries ever to grace creation. (I once asked my friend Marjan to mail me some overnight and she said, "I'd consider it if I wasn't certain it'd disappoint us both.") The rafters above were decorated with colorful beer taps and peculiarities like giant stuffed beings from Maurice

Sendak's *Where the Wild Things Are* (coincidently, my favorite childhood book). On the beams someone had written "the secret to life," which I always meant to write down in my small back-pocket notebook but never did. Moves always remind you of these tiny regrets.

For all these reasons, that small pub was the place chosen for my last night with the Idaho gang. A small gathering of people I had come to love saw me off. They bought me drinks and gave me gifts, letters, and apologies, and released debts. I paid back one couple the seventy-five dollars I still owed them for a washing machine and they ripped up the check in front of me. Inside jokes reigned. I accepted all their kindness with a straight face, but I knew as soon as I got back into the car, I would lose it.

Which I did. After a few minutes, I felt my face start to freeze from the tears (I had forgotten to start the car and get the heater going). The passenger-side seat was piled with presents and cards — so many that a person could not move them aside to sit down. As cluttered as it was, the station wagon had never felt this empty. I cried the entire ride home. Packing was not easy that night.

SO THIS WAS MY STATE OF MIND as I drove over that snowy pass: that awful pairing of fear and memories still too wet in my mind to paste into some cerebral scrapbook. I'd been pulled out of the state without warning, and was genuinely terrified about starting over in Vermont.

Knowing there was a roof waiting for me at the end of the highways gave me a taste of comfort. Somehow, I'd found a cabin to rent. That was pure luck — a seven-line ad in a local newspaper that I had raised an eyebrow at, then thoughtlessly thrown into my luggage the weekend I'd interviewed. When they offered me the job, I rummaged through the suitcase and resurrected the crumpled weekly. I called the number, all wrapped up in hopes that the place was still available. All I knew was it was affordable, allowed my dogs, and was only 12 miles from my new office. So when the kind Connecticut landlord told me I could shack up at her second home, I high-fived the air like a complete dork and hoped she'd allow chickens, too. I rented it sight unseen, arranging my new life over the phone. Staring out into the blizzard, I wasn't even sure I'd live to see it.

You're probably wondering why the exodus, considering how much I loved that farm and town. Sorry, kids, but sometimes love isn't enough. Thanks to issues with my old job (well, not an issue so much as the fact that it didn't exist anymore — 80 people at our company were laid off, thanks to the downturn in the economy, and I was one of them), I had to leave Idaho. There just weren't enough graphic design jobs with benefits to go around. I got lucky, though, and within a month of getting my pink slip, I'd landed a job with an outdoor retailer in Vermont that specialized in fly fishing. I was supposed to start this new job in less than two weeks. Everything I knew about fly fishing I'd learned from vaguely remembered scenes in *A River Runs Through It.* I didn't feel the need to share this fact with the sweater-clad outdoorsmen who interviewed me, however. Hopefully, I could pull this off.

COMFORT EVENTUALLY CAME that snowy night on the mountain. Even though I still couldn't see the brake lights of the truck in front of me, I'd made it through the worst of the pass and was finally heading down the other side. My spirits lifted and I started to focus on the road ahead. For my own amusement, I started up a conversation with the dogs. Jazz woke with one eye open, annoyed. (I didn't mind bothering him. I cover his rent.) I told my roommates what I knew about Vermont as I straightened out the wheels and got us back somewhere between the mountain and the drop-off.

"It's called the Green Mountain State. *Ver* means Green and *mont* means mountain. So it's not just a snappy nickname — it's literally called the Green Mountain. I really can't attest to this since I was there in January, but the hills looked like they *could be* green. Actually, the hills looked kind of like deflated volleyballs with angry toothpicks sticking out of them. Anyway, apparently they've got the mountain thing covered so we'll be okay."

I'd never lived away from mountains, and have grown to trust them like babysitters. In fact, I like to keep my area code around three things: mountains, trains, and crows. My corner of Vermont had all three. This, as ridiculous as it sounds, provided a thick blanket of relief as I drove east. I continued to pester my captive audience.

"I really don't know that much about these people, guys." The dogs yawned. "I do know they've got the sustainable farming thing down. It's a big agricultural state, even if it's small. Did you know the only state with fewer people than Vermont is Wyoming? How do you guys feel about that? I'm torn." Annie leaned her head against the glass and looked out the window as the flashing lights of the pass' entrance came into view. "Let's see, there will be hard cider, an amazing fall, and all the maple syrup we can swill . . . ."

Truth is, my facts were a little sketchy. I hadn't done a lot of research before the move. Growing up in the domesticity of northeastern Pennsylvania, I was always given the impression that Vermont was a place for feral New Englanders. A wild place of ski resorts, crunchy organic farmers, jam bands, liberal politics, and fervent college students. These were all things I enjoyed, but it never seemed like a realistic place to *actually live*.

I had a mild hunch I'd fit in in that corner of New England. Vermonters seemed like my kind of people. From goat dairies and cheese-making operations to organic orchards and cider mills, the place is an artisanal agriculture Mecca. I knew that a lot of writers lived there, too. In fact, I'd just read an article in *Vermont* magazine about all the writers who choose to live there — typing indoors while the snowstorms holed them in. It seemed homey, plausible. I'd give it my best. Perhaps I'd always been a Vermonter and simply paying taxes in the wrong states all along.

As good as Vermont sounded, had I known that the drive back East would've been so god-awful, I might've reconsidered. I hadn't been wrong when I noticed I was starting to feel ill. During that road trip, I had the worst case of the flu I have ever experienced. For four miserable days, I drove the 2,800 miles from Sandpoint, Idaho, to Sandgate, Vermont, leaving one mountainous, ice-covered place for another and hacking most of the way. Adding to the fun was a serious case of insomnia; so most days were spent dragging tired nausea across time zones. I threw up at Little Big Horn and felt guilty about it.

After four long days of popping Dayquil and memorizing truck stop logos, I pulled into my new driveway in Sandgate, Vermont, not knowing what lay in store. Slipping on the icy driveway, I scrambled up to the porch and, as my foot hit the faded green paint, I knew this was the right place for me. The cabin was small, but the potential huge. Under the ice, I could make out the fence around a large garden, and right across from it was an abandoned storage shed. Already my mind saw green vines from spring pea tendrils crawling along the icy, rusted, gate. I could hear my future hens clucking in the shed (which, I was certain, would make a fine coop).

Inside, the small log-cabin style camp house was perfect for the dogs and me. Barren, yes. It needed my letterpress posters from Tennessee and some of my more whimsical accoutrements. (Looking around after stepping inside, I found the perfect place for my hand-cranked Dancing Nixon music box, which is exactly what you think it is.) I knew with some time, some hard work, some chicks in a brooder box, and a warm bed, this place could be everything the Idaho farmhouse was and more. The pasture-sized lawn outside teased at the idea of hoofstock. Would this be the place I strung out rolls of fencing and finally found my first flock of sheep? The future was comatose in February icicles but still had a heartbeat under it all. I knew what this corner of the world could be. The dogs, in their own way, agreed, and curled up to sleep that night in a spot right in front of the fireplace. Dogs know.

OUTSIDE THE FARM

It's time to make some friends

SINCE I'VE STARTED OUT on this road to my own farm, I've met a lot of colorful characters along the way. There's my next-door neighbor Ron, who raises horses and tells me about his childhood ponies when he's fixing my sink. There's Emily (I met her at my Tuesday-night bluegrass jam); she plays a mean guitar and has been eating a strict local diet for the past year. There are the people at the farmers' market who sell herbs and pottery, like Merla, who made a little personalized garden plaque especially for me. These people are my soil. They are what enrich my experiences and bring life to the pages of my story.

Had I never made the effort to get to know others, I would have lost so much. I urge you to also go out and meet the people living the way you aspire to live. They have more to teach you than anything this or any other book ever could, and they'll be happy to help you. I know if someone e-mailed me and asked to help her get started with chickens or planting a small garden, I'd be on cloud nine. Someday I hope to be the woman who orders a box of chicks on a snowy day with five extras going in a shoebox to another home.

VISIT A LOCAL FARM

People who practice self-reliant living are some of the most interesting and welcoming people I've ever met. I think when anyone who created a farm meets someone who's just starting out, an ingrained mentor starts pulsing. Their eyes light up when they meet new people who want to swap gossip about the best compost turners or have some tips on gray-water drainage. If you have no idea what those things are, all the more reason to consider conversing with the experts.

Every time I've visited an off-the-grid home or small farm, I've felt as comfortable as a long-lost cousin. The folks who live there patiently answered my questions, corrected my mistakes, and shared advice and stories over meals and barn tours. And I say with complete confidence that I've retained more priceless information about self-reliant practices that came to me over kitchen tables than from any other source. There's something about learning new information when you're face-to-face with the person giving it (and the chicken he is talking about) that makes you remember. You need to make the effort to talk with people who are doing what you want to do. It's that simple. There's nothing to be shy about. Even if all you've done is read this book and start sprouting a half-dozen seeds on your windowsill, you have more than a year's worth of dinner-conversation starters with your local organic farmers. They'll happily tell you about what worked for them and what didn't, how high to hang the lights above the seedlings, what you can add to the soil to grow bigger veggies.

As contradictory as it may seem, I'd say start online. I don't recommend driving around the countryside and knocking on doors of houses with barns and horses out back. There are sites, such as Eat Well Guide, that list organic farms by zip or area code complete with phone numbers and e-mail addresses. If you're sincere about wanting a more sustainable life, no matter how you interpret that, visiting successful homesteads and farms is integral to your education. Send a simple note saying you're interested and just starting out but would love a short tour and a chance to stop by and see their setup. Most small operations are eager to share with wide-eyed beginners everything they've worked to accomplish. You'll find the hardest part of visiting strangers' farms is getting the nerve to ask the owners in the first place, but it's all downhill from there.

GO TO THE COUNTY FAIR

I entered my first county fair knowing absolutely nothing about poultry shows. I had a pair of Black Silkie Bantams in a cardboard box under my right arm. Someone showed me how to fill out the forms, how to hang the cage tags, and explained the care and feeding responsibilities of the exhibitors. After a whirlwind of feathers and paperwork, I had them entered in the small show. Those little black chicks I raised from babes were strutting and crowing for the whole county to see. I was genuinely proud of them. I didn't know the first thing about conformation of Japanese bantams, but they were bright-eyed and seemed happy. The night before, I had washed them in natural dish detergent and gently blow-dried them while we listened to the back-porch bluegrass radio show on public radio. The notion that a few years ago my design-school friends and I were skipping through Chelsea galleries and eating overpriced French toast at the Empire Diner made me laugh out loud while I towel-dried my pathetic wet hen's face. Unlike many of those old college friends, I had yet to travel overseas, but I still felt I had made one hell of a journey.

The fair lasted a whole week. Before work I would stop by and make sure my birds had plenty of food and water. I ended up not only winning a few ribbons, but also taking home the grand champion rooster in the process! A couple of us chicken people were standing around the poultry barn, squawking louder than the hens about our birds, and before we knew it we had a trade going on. She would take home my Black Silkie rooster and I would take her Welsummer named William. Which fit him perfectly. I knew no chicken that looked half as regal. He had a giant green plumed tail and a big red crown. Did the people at Kellogg's know their spokesman was in Idaho? I brought him home and within minutes Mary Todd Lincoln found her soul mate. So if nothing else, I made a damn good chicken matchmaker.

BECOME AN AGRITOURIST

If the idea of cold-calling a random small farm makes you uncomfortable, you have other options. Agritourism might be the perfect plan B for the socially faint of heart. The idea is to make small farms more accessible to people who are interested in seeing the process close up. Everyone from apple growers to ostrich farmers is opening the gates to the public. In my small town alone there are alpaca farms, you-pick orchards, wineries, horse ranches, and tree farms all welcoming the community to change up their usual weekend activities and stop by and see them. Agritourism is becoming more and more popular all over the country. Bed-and-breakfasts, group tours, and even beginner's classes are forming in every state to get people like us involved in some interactive classes. This could be a perfect first step for you. Taking a few trail rides at a tourist ranch might give you enough courage to talk to a rancher at a smaller operation and see how he or she got started. If nothing else, you'll have a few, but poignant, shared experiences to start with.

JOIN A HIKING CLUB

While hiking clubs aren't exactly the greatest option for learning how to compost properly, they do seem to be a hotbed of like-minded people. There is something about the mentality of the outdoor enthusiast that craves self-reliance. Most people who choose to spend their time outside have a lot of common values about conservation. I've gone on group hikes and found musicians, preachers, teachers, and students, and all of them had great advice on the best time to pick sweet corn (there were heated silk-tassel debates). If your city doesn't have a hiking club, try joining a national organization like the Sierra Club. You'll find local chapter activities and how to sign up for volunteer work all over the world if you're into that kind of thing. The memories you could make spending a summer clearing hiking trails in Maine would be well worth the paperwork.

SIGN UP FOR A CLASS

Colleges and community centers offer classes in many of the topics discussed in this book. Yarn and craft stores offer knitting and quilting classes. Colleges host lectures on solar-panel water heaters and your town hall might have a speaker with advice about following a smarter recycling program. Pay attention to those bulletin boards at the natural foods store and the back of the weekly reader. You'll find that all sorts of people are taking Shepherding 101 and learning how to bake perfect cookies. Sign up and go check it out — you never know who you'll meet. Chances are someone will click with you when you're elbow-deep in batter and laughing, and if not, you still went ahead and took positive steps toward doing something new.

SHOP AT THE FARMERS' MARKET

I keep talking about farmers' markets. That's because I cannot stress enough how important they are as a resource to us beginners. The people behind those folding tables and sunflowers were either in the exact same place you are now or grew up knowing a manure spreader was a piece of farming equipment and not the town gossip. Either way, they are invaluable Research, Son. Next time you come across an open-air market and the people seem genuine, walk up and say you really want to get into their line of work. Even if you explain it's just a few container gardens on New York's Upper East Side, they'll be glad to know a fellow gardener and appreciate the friendly conversation. Don't be shy: tell them about your eggplants, and if you have a question about why your lettuce gets slimy at the base or why your pumpkins aren't coming in, ask them. They'll be the closest thing to a house call you can get.

WANT MORE?

Recipes and projects (overalls optional)

SO YOU'VE HEARD my story so far — how I went from a hopeless beginner to a farm girl in her own right. It was a heck of a ride, but I didn't get there without some serious trial and error (which my smoke alarm can attest to. Don't forget that loaf in the oven, folks . . .). While it was realized with the help of wonderful people and indispensable advice, a huge part of it was just diving in and trying some of this stuff on my own. Now it's your turn to grab a sewing kit or some mixing bowls and start creating.

In this section, I'm leaving you with some recipes, projects, and patterns to make everything from my dad's favorite apple cake to the canvas tote bag you can bring it to work in. You'll learn things one bowl of batter or sewing project at a time, so have fun with them and don't take them too seriously. The point is to enjoy yourself every stitch of the way.

RECIPES

STRAWBERRY JAM

Makes about 5 half-pints

2 quarts fresh strawberries, cleaned, stemmed, and coarsely chopped

7 cups sugar

1 packet pectin

Lemon juice

Canner or large saucepan with lid

Canning tongs

Canning rack

5 half-pint jam jars with sealable lids

Large wooden spoon

Plastic kitchen funnel

*You don't need any complicated supplies to make jams — it can be done with a large pot right on the kitchen range. If you choose to can some, a few handy canning tools like jar tongs (for pulling steaming hot jars from the water bath) and a small funnel (for cleaner pouring) will make the job a lot easier. If you'd rather not mess around with canning, you could make freezer jam (**see* Freezer Jam, *page 164**), but this makes it harder to give as a gift.*

The first time I made a batch of jam, I did it in my old dented spaghetti saucepan and the jam came out perfect. One word of caution: You can't can everything using the method I describe below.

If you want to put up veggies from the garden, it's a good idea to consult a canning book (see Research, Son, page 185) so you don't give all your friends botulism. Fruit preserves are so high in acid, you won't need to worry about this.

Sanitize the jars. To start, you'll need to sanitize your jars. Fill the canner or stockpot halfway with water, set it on the stove, and bring it to a boil. While you're waiting, wash the jars and lids with hot, soapy water, then rinse them thoroughly. Once the water is boiling, put the washed jars and lids into the bath. (Make sure the jars are still warm from being washed; if you drop ice-cold jars into boiling water, they'll crack.) After a few minutes, take them out with the canning tongs and set them face up on a clean towel, ready to be filled with jam.

Making the jam. Pour the berries into a saucepan and bring them to a boil over medium heat, stirring occasionally with the wooden spoon. When it starts to boil, add 7 cups of sugar and keep stirring. Slowly add the pectin, stirring it in completely. Keep stirring and bring the jam up to a rolling boil. Let it boil for a full minute, feeling the wooden spoon work harder and harder as the jam thickens. Turn down the heat to a simmer while you get ready to fill the jars.

Seal and store. Pour the jam into the jars, using the funnel and trying to keep the rims as clean as possible. Leave about ½ inch between the jam and the top of the jar (called "headspace") to create a vacuum and seal the lids. Wipe the rims of the jars with a clean, damp cloth; place the lids; and screw on the rings almost all the way.

Place the canning rack inside the pot or canner and get the water back up to a boil. When it's going strong, use the tongs to gently place the closed jars of jam into the canning rack in the water bath. Put the lid on the canner and let the jars boil along for 12 to 14 minutes. At this point, remove the jars and set them on a clean towel on a level countertop. Do not disturb them for at least 12 hours (I like canning at night, so that in the morning the jars have set and are ready to be stored or given away). You'll know the jars have sealed correctly if the lids are slightly concave and do not budge or pop up if pressed. If you have any jars that did not seal tight, put them in the fridge or freezer immediately for storage. The rest can be dated and stored in a cool, dry place.

When you open up your home-canned jam, use common sense before you spread it on your toast. Does it have a loose lid? Does it look and smell just like it did the day you made it? Is there any mold or weird coloring? Canning is a great way to preserve food, but it must be practiced with caution. If you're not 100 percent sure you can pull it off by yourself, ask an experienced canner to walk you through your first batch. She'll probably be thrilled someone wants to learn, and may even let you borrow her supplies. Make sure you send her home with a jar or two for her trouble.

FREEZER JAM

If water-bath canning sounds a little too complicated for a first-time jammer, you can always opt for freezer jam. Mash 4 cups of berries in a large mixing bowl, then stir in 1 tablespoon of lemon juice and set aside. Pour 1 cup of water into a small saucepan and mix in the pectin and 3 cups of sugar, stirring well. Bring the solution to a boil and let it boil for a full minute, stirring constantly. Pour the solution over the berries and mix it in well. Pour the berry mixture into sterilized jars, filling them only ¾ full — the jam will expand as it freezes, and will crack the jars if they're filled completely, so give it some breathing room. Let the jam set completely (about 24 hours) at room temperature before storing jars in the freezer where they will keep for up to a year. When you take a new jar of jam out of the freezer, let it thaw in the fridge. It'll keep for about 3 weeks and will taste even fresher than the canned stuff. Nice.

SIMPLE HOMEMADE PASTA

Serves two

3 large eggs

2 to 3 cups all-purpose flour

My friend Diana taught me how to make pasta over at her family's farmhouse. When we had rolled, sliced, and set out all the long, skinny strands of noodles to be cooked, it seemed like too little for dinner. I raised an eyebrow at this and asked if she thought it would be enough to feed three adults. She assured me it was, that real pasta (as she called it) was thicker and more filling than the dried, boxed kind. She was dead-on. After a few bites of the thick, juicy noodles and homemade chunky sauce, I felt like I had eaten three times as much as what was on my plate.

Crack the eggs into a medium-size bowl and beat them lightly. Add flour to the eggs, ½ cup at a time, mixing until you have a slightly sticky dough. Knead it for a short time and when you're content with the ball, wrap it in foil or plastic wrap and let it rest in the fridge for an hour.

After it's had its nap, place it on a lightly floured surface and roll it as thin as you can while keeping it a solid sheet of dough. Sprinkle flour over the sheet of dough and gently roll it up into a long tube. Then cut the tube into ⅛- or ¼-inch-thick slices and unroll each slice into a long strand of pasta. Drape them over a coat hanger as you unwind them; this keeps them from sticking together before you're ready to boil them.

When all the noodles have been hung, heat up a large pan full of water until it boils. When the water is bubbling, toss in the noodles and let them cook for 3 to 5 minutes. It takes much less time to cook than dried pasta does. Drain and serve with Chunky Pasta Sauce, below.

CHUNKY PASTA SAUCE

Serves four

8 large or 10 to 12 small fresh tomatoes
(canned will work, too, if it's off-season)
2 to 3 tablespoons olive oil
¼ to ½ cup minced onion
4 garlic cloves, peeled and diced
½ cup fresh mushrooms
¼ cup fresh basil (add more if you're into it)
At least 1 teaspoon garlic salt
At least 1 teaspoon dried oregano
At least 1 tablespoon Italian seasoning

Here's the companion to homemade pasta. I learned the recipe from my friend Bruce (Diana's husband). He helped me cook and can eleven jars of this sauce from the giant tomato crop my garden coughed up. Let me tell you, it's really nice to be dining on fruits from the garden when there's seven inches of snow on the ground.

Mash the tomatoes with a potato masher or in a food processor, then drain off about half the liquid and set aside the tomatoes in a bowl. Put a large saucepan on the stove over medium-high heat and coat the bottom with olive oil. When the oil has heated up to a bit of a jumpy state, add the minced onion and sauté until it's translucent. Add the garlic and sauté briefly. This base is what your sauce will be built on.

Pour in the tomatoes and stir the concoction with a wooden spoon. Add the mushrooms and a healthy dose of basil, garlic salt, oregano, and Italian seasoning. Suggested quantities are listed above, but tomato sauce is a breathing recipe that you should adjust to your own taste buds. If there's something I didn't mention that you love in sauce, like chilies or cooked sausage, throw it in, too.

When it's all mixed in, cover it with a lid and turn the heat down to an even, low temperature, so the sauce is barely bubbling. If you have the time, let it simmer for a few hours to "get to know itself," as Bruce would say. Stir it every so often, so it doesn't burn or stick. When you've sampled it and it tastes like it was born to cover pasta, you've got yourself a fine sauce. Bake up some garlic bread sticks and grab a bottle of local white wine, and enjoy your completely farm-made meal.

DIY PIECRUST

Makes two 9-inch piecrusts

1½ sticks of butter, chilled
2 tablespoons vegetable shortening
1 tablespoon sugar
2 cups flour
3 tablespoons ice water (in a cup with ice cubes)

Although I tend to use frozen piecrusts for my quiche (because they're handy and I usually have at least one in the freezer), you can make them from scratch in advance and store them in the fridge if you like. Making crust is simple, and after all, if you don't make the crust yourself, the quiche isn't technically "homemade," now, is it?

Dice the butter and shortening into small cubes for mixing. Combine butter, shortening, sugar, and flour in a bowl or food processor, until the mixture has the consistency of cornmeal. Add the ice water and mix it in just until the dough comes together. Split the dough in half, pat each half into the shape of a disk, and wrap each in waxed paper. Let rest for at least an hour in the fridge, or up to two days, before using. Each half makes a 9-inch crust when rolled out.

Quiche crusts need to be "baked blind," as the pros say — this just means you need to bake them before you put any filling in them. If you've got two pie pans, you can bake both piecrusts, use one, and put the other one in the freezer for a future quiche-crust emergency.

When you're ready to roll, preheat the oven to 350°F. Roll out each disk of dough on a lightly floured surface and lay each one gently into its own pie pan. Trim the edges, if needed, but leave a little overhang — crusts that are baked blind tend to have a lot of shrinkage. Fill the crusts with dried beans (or pie weights, if you've got 'em) and bake for 15 to 20 minutes, or until the crusts are lightly browned. Let them cool and remove the beans. Slip one crust (still in the pan) into a ziplock bag and put it in the freezer. Use the other crust for Three-Hen Quiche.

THREE-HEN QUICHE

Makes one quiche

Olive oil
⅓ cup diced onions
1 broccoli crown, chopped into 1-inch pieces
3 large eggs
1 cup milk
Salt and pepper, to taste
1 piecrust (frozen or homemade — see DIY Piecrust, page 168)
½ to 1 cup grated Cheddar and Monterey Jack cheese

Over my girls' coop, there's a sign that reads Team Quiche. *That's because my three original hens would lay an egg a day each, and three eggs are exactly what my favorite recipe calls for. Quiche is easy to make, great to share, and keeps well in the fridge. It tastes just as good heated up in the oven the next day. You can customize it, adding whatever you prefer. If you use a breakfast meat like bacon or sausage, make sure it's cooked before you mix it with the other ingredients.*

Preheat the oven to 350°F. Heat the olive oil in a skillet on medium-high heat. Sauté the onions until they are browned and fragrant, then put them aside in a small bowl. Pour about ½ cup of water into the skillet and steam the broccoli over medium-high heat. When the broccoli is tender, set it aside in a small bowl.

Beat the eggs together with the cup of milk. Add a little salt and pepper to taste. Toss the onions and broccoli into the piecrust and pour the egg mixture over them. Grate the cheese over the top of it all. Bake, uncovered, for 45 to 50 minutes. Poke the tip of a knife into the center of the pie to check for doneness; if it comes out clean, without egg on it, you've got yourself a quiche.

JACKAPPLE CAKE

Makes one cake

3 large apples, peeled and diced*
1½ cups sugar
1 teaspoon baking powder
2 to 3 teaspoons cinnamon
¼ cup honey, heated
¼ cup fresh-pressed cider
½ stick butter, softened
3 large farm-fresh eggs
1 cup vegetable oil
1 teaspoon vanilla extract
2 cups all-purpose flour

Topping

⅓ stick butter, melted
¼ cup sugar
1 teaspoon cinnamon

*I like Braeburn or Gala; if you use a smaller variety, like Fuji, you'll need to use four. Do not use Red Delicious — apples, like people, can't be judged by how pretty they are on the outside. The suggested varieties aren't lookers, but they're damn tasty baked with cinnamon.

When I was growing up, in Palmerton, Pennsylvania, my dad made this recipe for us every autumn. Like turning leaves and school starting, my dad's apple cake was a sign of fall. He learned it from a friend, and since then we've shared the recipe with many others. I'm happy now to share it with you. Nothing tastes better with this than a mug of hot cider, and that's coming from someone who would rather have hot coffee during childbirth than an epidural.

Preheat the oven to 350°F. In a large bowl, mix the apples with the sugar, baking powder, cinnamon, and warm honey. Set the mixture in the fridge for 2 hours. When the apples have absorbed all the sugar and spice flavors, add the wet ingredients (cider, butter, eggs, oil, vanilla) and mix well. Add the flour, ½ cup at a time, and mix well after each addition. Pour the mixture into a greased 9-by-9-inch cake pan.

To make the topping, combine the butter, sugar, and cinnamon into an easily spreadable paste. Use a pastry brush to slather it over the batter, making a sugar crust that will bake into the top of the cake. Bake for 30 to 40 minutes. Check after 25 minutes. When a knife or toothpick comes out clean, it's done.

PATTERNS AND PROJECTS

MINI GREENHOUSES FROM TRASH

If you want complete control over the quality of your vegetables, not to mention a better selection of plants, starting from organic seeds is the way to go. You can buy all sorts of heating pads and seed-starter systems, and most of them are worth it, but I'll share my cheaper method that's just as effective.

All you need is an old soda can, a twenty-ounce soda bottle, and a desk lamp that takes a standard lightbulb. Also pick up some peat disks and soil-based peat pots. You can get these little brown wafers from any garden-supply catalog and most hardware stores. When warm water is added to the disks, they blow up into fertile soil cradles for your seeds. Cover the seeds with a little bit of that soil and place them in the matching brown peat pots. If you cut a rinsed-out soda can in half with sharp scissors, you've got two mini planters. Place a peat pot inside the half can and you've got a perfect little birthplace for your future harvest.

Now, take that soda bottle and cut off the neck, so the girth of the plastic is open at its widest point. Place the plastic dome over your can, and you've just created a mini greenhouse that recycles the trash. Set up the contraption under a tall desk lamp that's been fitted with a fluorescent bulb, and you're all set for heat and "sunlight." The moist soil, UV rays, and metal can create a warm and safe place to start even the most finicky seeds. The only real trick to remember is not to place the lamp too far from the seedlings once they sprout. If the plants grow too fast (to reach the light), they won't be stout enough to take real weather conditions when they move outside. You're aiming for fat little plants. When they grow too big to be enclosed in the soda bottle, you're ready to start gardening!

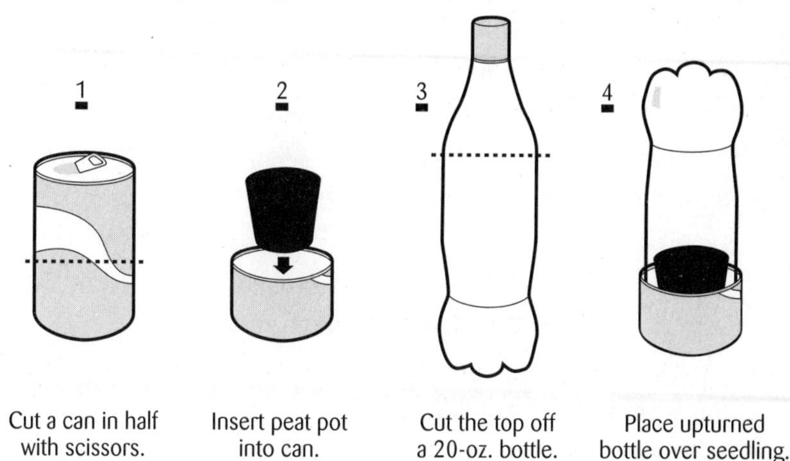

SEW YOUR OWN GROCERY BAGS

Some folks might think a sewn-up square with handles isn't a big deal, but if you've never sewn anything before, you'll be so thrilled by this simple project, you're going to be buying people drinks. Once you have the basic sewing kit collected, you can make your own shopping bags quickly and easily. This system has two parts: the large bag for groceries and smaller, lighter bags for bulk grains and produce. Those little plastic baggies you use for carrying peaches home are put to work for a few hours and won't decompose for over 100,000 years. Your old T-shirts, however, are already decomposing in that bottom drawer, so why not recycle them into a green alternative to plastic bags?

Take a trip to the fabric store and find the perfect canvas, duck, corduroy, or some other, heavier cloth fabric for the main bag. Choose something that can handle a full load of groceries but isn't too thick to run through a machine or hand-sew. Also make sure to pick up a yard of heavy-duty nylon webbing for the handles. Contrasting colors look great, so if you're buying a plain brown canvas, why not get baby blue webbing? This bag is not only going to be helping the environment, it will also give you the opportunity to make something you can be proud of and be creative with. You'll also want to grab about a yard of nylon cord for the ties on the T-shirt produce bags.

RUNNING WITH STITCHES

If you don't own a sewing machine and want to try sewing a few simple projects by hand, here are the three basic stitches you'll need to know. Keep in mind that sewing by hand does take longer than by machine, and the stitches aren't as sturdy. So stick to small projects and save the hard-core patterns for when you've acquired that vintage Singer.

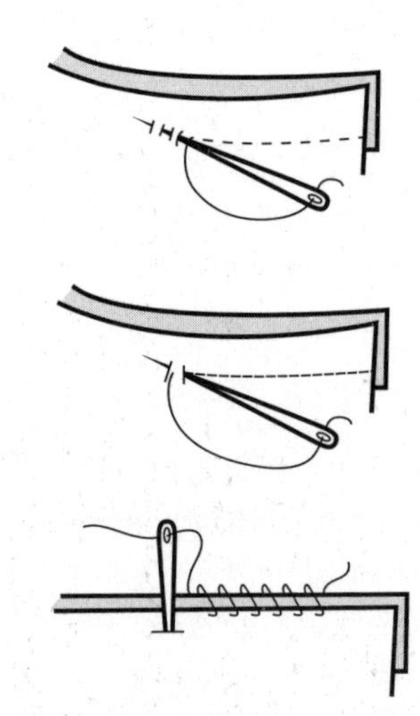

Running stitch. This basic stitch is what quilters use for hand-quilting.

Backstitch. Sturdier than the running stitch, this is the one to use for sewing durable seams.

Whipstitch. The whipstitch is often used to sew together two finished edges.

THE FABRIC GROCERY BAG

Fold the fabric in half and cut it into a square or rectangle of your desired size (I use a 26-inch square). You're folding it in half to ensure that you have two identically cut pieces, and I suggest making the fold in the fabric the bottom of your bag. It'll be stronger than a seam and will save you some time. Now you're ready to sew it together. Make sure you're sewing it with the wrong sides out ("wrong" meaning unprinted or underside). You're going to turn it inside out, so the stitches will be hidden when you're done. If you're using a needle and thread, stitch up along the sides from bottom to top with a tapestry needle and heavy thread. If you're using a machine, make sure the foot is high enough for the fabric to zoom through. When the sides are sewn up, knot off the thread. Now cut two equal-length handles from the webbing (half a yard always works; it's enough to sling over your shoulder) and attach them to the upper inside edges with a reinforced stitch. When the handles are sewn, you're ready to turn the bag inside out and take it to the farmers' market.

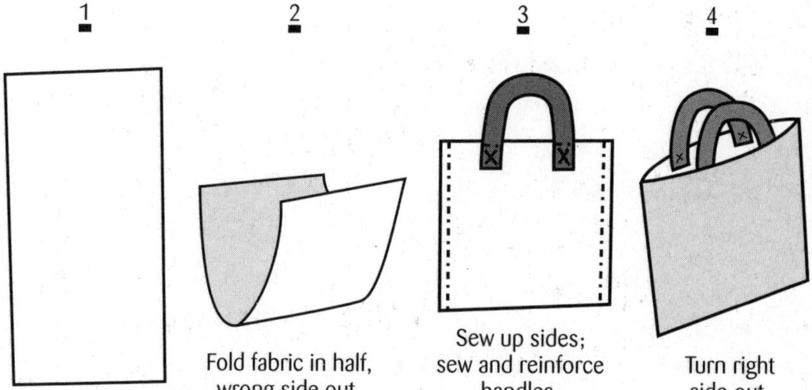

1

2

Fold fabric in half, wrong side out.

3

Sew up sides; sew and reinforce handles.

4

Turn right side out.

THE T-SHIRT PRODUCE BAG

Cut off the sleeves and neck of the shirt, so all you have is a fabric tube. Turn the shirt inside out and sew the bottom edge shut. There you have it. All you need to do is cut a foot or so of cord, fold it in half to find the center point, and sew that point about 3 inches from the top of your new cloth sack. Now you have a tie built in to keep your tomatoes or oatmeal in line while you're shopping.

Cut T-shirt as indicated.

Turn inside out and sew across bottom and top.

Turn right side out, insert cord in top flap, and cinch it up.

EASY FLEECE MITTENS

To make your own fleece mittens, buy a yard of winter fleece in your favorite color at the fabric store, matching thread, and a felt-tip marker. Fold the yard in half and place your hand on it with your fingers slightly separated (so your mittens won't be too tight). Trace around your hand, leaving a halo of about ¾ inch, and let the base of the mitten come to about 2 inches below the top of your wrist. Trace it again, and when you have two outlines on the folded yard, pin the fabric together inside the pen lines and cut out the gloves. Turn the top two pieces top side down, so all four "hands" are now wrong side up.

For each piece, fold the wrist toward you about ½ inch, then sew each fold in place with a running stitch right down the middle. Pin two "hands" together with the edges matching and the folds facing out. Sew them together about ¼ inch from the raw (unstitched) edges, going from one end of the cuff and around the other. Turn the mitten inside out, and there you have it! Do the same thing with the other two "hands." Make them special by embellishing them with patches or other adornments.

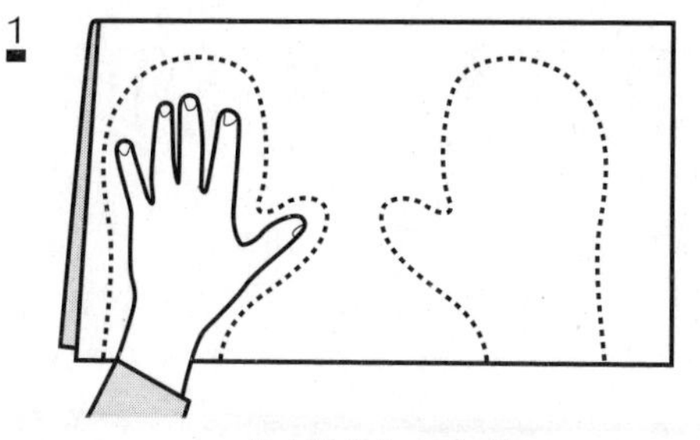

Fold fabric in half (wrong side out) and trace left hand, then right hand.

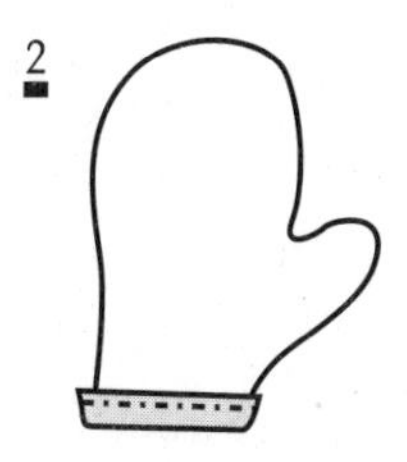

Turn up bottom to make a cuff and sew across. Repeat with other 3 "hands."

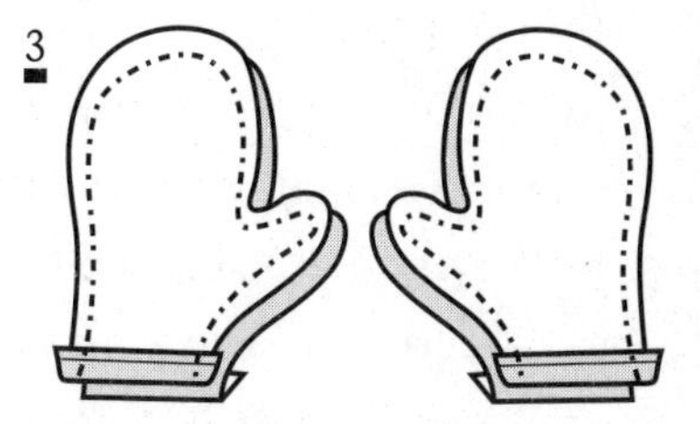

Place in 2 pairs, and sew along outside edge.

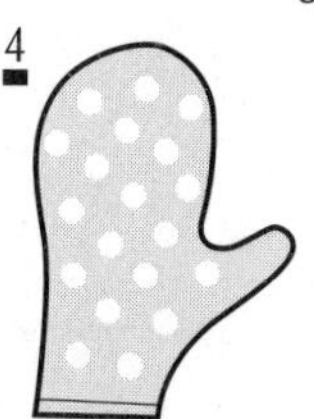

Turn right side out.

RESEARCH, SON

OKAY, SO THE BOOK IS OVER, but your research is just beginning. There's a good chance you can access everything I list below at your local library. If they don't have it, they'll find it and have it for you as soon as they can. Besides required reading, you'll find Web sites and catalogs that offer everything from butter churns to modern-style igloo chicken coops. I hope all this helps you reach out to the self-reliant community one click, phone call, or page at a time.

GENERAL INTEREST

Cold Antler Farm

My farm's blog. It's a place for notes, pictures, drawings, and ramblings. I post songs and essays, sometimes a recipe or two. If you're interested in what's going on at my place, or what Jazz and Annie are up to, check it out. » *www.coldantlerfarm.blogspot.com*

—

The Contrary Farmer by Gene Logsdon

Everything by Gene is golden. His books are all entertaining, informative, and, in some cases, life changing. This one is the first book of his I ever read. It was given to me as a gift and has been a constant source of inspiration.

—

Better Off by Eric Brende

A memoir about a year without technology, written by an MIT grad student who takes his new wife and lives in a primitive Amish community for a year. Eye-opening stuff.

—

Encyclopedia of Country Living by Carla Emery

Every home should have one of these on hand, country or not. The biggest, most comprehensive work on general homesteading ever written. It's gone through a bunch of revisions, and the newest version even has Web sites and e-mail addresses.

—

MaryJane's Idea Book by MaryJane Butters

MaryJane is the queen bee herself. Her gorgeous idea book (cookbook, life book) is full of photography, musings, plans, suggestions, recipes, stories, and, of course, ideas.

—

Hit by a Farm by Catherine Friend

A hilarious and educational memoir of two ordinary women who take on a flock of sheep and end up true-blue shepherds.

—

Storey's Basic Country Skills by John and Martha Storey

One of the first books I ever really dove into when I started scaling down. Great reference and has everything from recipes for homemade cheeses to how to build a cowshed. It's aces, kids.

HOMESTEADING

The Good Life by Scott Nearing

The modern homesteader's bible. This is the book that inspired thousands of people to seek a self-sufficient life. A must-read.

—

Countryside & Small Stock Journal magazine

Over ninety years in publication and still going strong. *Countryside* is the magazine for modern homesteaders, with most of its contributors being readers with good advice. You can pick it up at any large bookstore, and every issue is worth saving for reference somewhere down the line. » *www.countrysidemag.com*

—

Urban Homesteading

Online resource for all things self-reliant. Specializes in city and suburban homesteading. » *www.urban-homesteading.com*

—

How to Homestead

A cool little Web site that's full of videos and Research, Son for future homesteaders, with a heavy focus on the urban scene. » *www.howtohomestead.org*

—

Homegrown Evolution

The blog and online home of Erik Knutzen and Kelly Coyne, authors of *The Urban Homestead.* Articles, wit, and reviews from a pair of city farmers who've done possibly everything there is to do under streetlights! » *www.homegrownevolution.com*

—

Lehman's

The best nonelectric catalog out there. This Ohio-based company supplies many Amish homes across America, as well as modern homesteaders. From garden gear to hand-cranked washing machines, these guys have it all if you want to go back in time. » *www.lehmans.com*

—

Hobby Farms magazine

One of the reasons I got into this mess in the first place. It was thumbing through issues of *Hobby Farms* in college that jolted me into my hunt for my own little farm. A great resource all around.
» *www.hobbyfarms.com*

—

Barnyard in Your Backyard edited by Gail Damerow

Covers beginner info on all sorts of animals in your future menagerie. Practical, much-needed advice on cows, chickens, sheep, geese, ducks, rabbits, goats, and more. Information with the hobby farmer in mind.

—

The Have-More Plan by Ed and Carolyn Robinson

This awesome book from the 1950s started many people on their own paths to self-sufficiency. It's jam-packed with info, is easy to read and understand, and was one of the original small-farm start-up books of its kind.

GREEN LIVING AND RENEWABLE ENERGY

Green People

Online hub for green living, news, and products. A large directory of companies and online communities. » *www.greenpeople.org*

–

Home Power magazine

A periodical for family and small-scale green energy operations. Every issue is loaded with advice, phone numbers, local festival connections, and everything else you could use to start soon. » *www.homepower.com*

–

Carbon Fund

Where you can learn about offsetting your carbon emissions, green energy, and how to get more involved in renewable energy. » *www.carbonfund.org*

–

Gaiam

A catalog that specializes in natural and earth-conscious living. These folks carry everything from all-natural toilet paper to organic-cotton pajamas. » *www.gaiam.com*

–

Treehugger

True to its name, Treehugger is a green-living blog for modern eco-America. It is updated so often it's staggering, and it lists the latest news, reviews, and trends in the world of renewable energy and conscious consumerism. » *www.treehugger.com*

GARDENING

Seeds of Change

Organic seed and gardening supplier. Beautiful online catalog and great selection. It also supplies things like kitchen-counter composters and organic-cotton garden wear. » *www.seedsofchange.com*

–

Pinetree Garden Seeds

Maine-based seed company with a large selection of heirloom and hard-to-find varieties. Four words: 'Moon and Stars' watermelon. » *www.superseeds.com*

–

You Grow Girl

Book, blog, news, and community for new and urban gardeners. » *www.yougrowgirl.com*

–

MaryJanesFarm

MaryJane's Idaho-based farm holds classes on farming skills and offers a great online community of farm gals that's alive and kicking. » *www.maryjanesfarm.com*

–

Food Not Lawns by H. C. Flores

An in-depth guidebook for turning your lawn into a garden and your neighborhood into a community. A vast collection of tips and ideas for anyone thinking about adding some green to his or her life.

–

The Vegetable Gardener's Bible by Edward C. Smith

Full-color reference and instruction book that everyone should have handy, whether you have three acres of veggies or a pot of peas on the windowsill.

–

The Gardener's A–Z Guide to Growing Organic Food
by Tanya L.K. Denckla

Conversational and comfortable reading for even for the newest gardener. A great first-timer's guide that'll have you planting like the pros and organically controlling diseases and pests.

—

The Veggie Gardener's Answer Book by Barbara Ellis

Quick answers to gardeners' constant questions. Handy, easy to read, and small enough to take with you to the garden.

KITCHEN AID

Take Two & Butter 'Em While They're Hot by Barbara Swell

An Appalachian collection of heirloom recipes and kitchen wisdom. This book, and many wonderful others, is available on the author's site, Native Ground Music. » *www.nativeground.com*

—

Ball Canning

The company that brought you the famous jar also has a site with free recipes for preserving fruits and veggies. Also lots of canning goods and supplies. » *www.freshpreserving.com*

—

Lodge Manufacturing

The best. A cast-iron-cookware supplier if you're in the market. These guys make everything from the basic skillet to outdoor fire-pit Dutch ovens. » *www.lodgemfg.com*

—

Blue Ribbon Recipes by the Farmers Almanac

Prizewinning recipes from a collection of country homes. Don't have a kitchen without it. » *http://store.almanac.com*

—

Eat Well Guide

A Web site dedicated to helping you find high-quality, healthy restaurants and local small farms. » *www.eatwellguide.org*

—

The Big Book of Preserving the Harvest by Carol W. Costenbader

If you have a garden and if you like the idea of still enjoying it in January, get this book, buy some Mason jars, and clear a September weekend on your calendar.

—

Serving Up the Harvest by Andrea Chesman

You grew it, might as well learn how to prepare it in a delicious way!

WORKING DOGS

Black Ice Kennels

A Minnesota kennel and outfitter that specializes in working pack and carting dogs. Harnesses, carts, packs, and more.
» *www.blackicedogsledding.com*

—

Wolf Packs

The best dog packs I've ever used. A small company in Oregon that makes a great fitted product for dogs of all sizes.
» *www.wolfpacks.com*

—

Ruff Wear

Canine outfitter that specializes in packs, travel gear, and working dogs.
» *www.ruffwear.com*

—

Wilczek Woodworks

Cart and harness supplier. Beautiful handcrafted work. Makes a great starter set for training. » *www.wilczekwoodworks.com*

—

Sled Dog Central

The largest mushing community site online, with long lists of outfitters, mentor programs, breeders, and working dogs in your area. » *www.sleddogcentral.com*

CHICKENS

Backyard Chickens

Online community and Research, Son for small-scale poultry farmers in urban and suburban areas and links for homemade coop designs, breed finders, and local suppliers. » *www.backyardchickens.com*

—

My Pet Chicken

A great online supplier for your beginning backyard poultry operation. What sets it apart is that it mails orders of chicks as small as three birds. It also has coops and plans, feed, fencing, and more. » *www.mypetchicken.com*

—

Murray McMurray Hatchery

My favorite hatchery. A large selection of chicken breeds for mail delivery right to your door. Also supplies everything you could ever need for farm and coop. » *www.mcmurrayhatchery.com*

—

Living with Chickens by Jay Rossier

An amazing book and resource for the new bird lover. Possibly the best modern manual on backyard poultry available today. Lots of great photographs, too.

—

Keep Chickens! by Barbara Kilarski

A must-have for any urban flock owner. Written by a Portland lady who invited a few hens into her life and it all went uphill from there.

—

Storey's Illustrated Guide to Poultry Breeds by Carol Ekarius
Amazing photography, encyclopedia-like feel, and plenty of breed-specific information, tips, and heritage history. A beautiful, fun book.

—

Storey's Guide to Raising Chickens by Gail Damerow
A must-have addition to the chicken library. Everything you could possibly need to know, even the gross stuff.

—

Chicken Coops by Judy Pangman
Plans, pictures, and inspiration for anyone with a backyard flock or a full-out farmstead. If you're building from scratch, get this book.

—

Omlet
Makers of the Eglu, the coolest backyard coop available to the public today. A little pricey, these run you about the same as a new iPod, but I assure you they are much more fun. » *www.omlet.us*

RABBITS

Storey's Guide to Raising Rabbits by Bob Bennett
Everything you need to know to own, breed, and raise bunnies.

BEES

Betterbee

New England beekeeper's heaven. » *www.betterbee.com*

–

My Beehives

Free online tool to track and keep a digital log of your hive. Also has message boards. » *www.mybeehives.com*

–

Dadant and Sons

Leading supplier of gear for all levels of hobbyists.
» *www.dadant.com*

–

Western Bee Supplies

Montana based, great service, and if you live on that side of the Mississippi, quicker shipping. » *www.westernbee.com*

SEWING AND KNITTING

Built by Wendy

A Brooklyn-based seamstress whose clothes I adore. Even better than that, she is the author of a book that inspires people to make their own wardrobes and it comes with three patterns. It's a fun introduction to sewing, and Wendy's definitely got the coolest, most modern patterns on the market today. » *www.builtbywendy.com*

–

Simplicity

A pattern collective. If you want to make something by hand, Simplicity has an inexpensive pattern to show you how.
» *www.simplicity.com*

–

Stephanie Pearl-McPhee

Author, blogger, and, more important, knitter. Has a great site with her inspirational fiberland updates, tips, and links.
» *www.yarnharlot.com*

—

Knit Picks

An online retailer that carries yarn, needles, and patterns you can buy as sets depending on the project. It also offers free downloads of said patterns, which is wonderful if you already have a few skeins lying around with nowhere to go. » *www.knitpicks.com*

—

Sew What! Skirts by Francesca DenHartog
and Carole Ann Camp

Get started in sewing your own skirts (sorry, guys . . . kilts?) today. The book uses pattern-free designs based on self-measuring, for a perfect fit every time.

—

Sew What! Fleece by Carol Jessop and Chaila Sekora

Just as simple and intuitive as the skirt book, but has more cozy ideas for everything from hats and gloves to full-out frock-style jackets. Many things you can whip out in under an hour. (Did someone mention Christmas gifts?)

MUSIC AND JAMS

Native Ground

Wonderful beginners' instructional books and recordings. When you've learned all there is in those, it has equally wonderful intermediate and advanced books and recordings! It also publishes books of pioneer lore, log cabin recipes, old-time music, and advice books. A gem.
» *www.nativeground.com*

—

Patrick Costello

Author and blogger who has a witty and logical approach to teaching old-time banjo through his book *The How and Tao of Old Time Banjo.* The book comes with a CD, as well as sayings and stories that help you feel like you're part of the process. » *www.funkyseagull.com*

—

Wood 'n' Strings

Tennessee dulcimers and mountain-music wares. Handmade instruments by Mike Clemmer, as well as fiddles, mandolins, hammered dulcimers, psalteries, and anything else you'd hear in the hollers. » *www.clemmerdulcimer.com*

—

Acoustic Friends

Social networking site for acoustic musicians. Think MySpace for the string-playing set. A place to find local jams, make friends, and see who else is picking in your zip code. » *www.acousticfriends.com*

—

Banjo Hut

A Knoxville-based banjo supplier that sells good-quality complete beginner packages of set-up five-string banjos. Its kits include tuned banjos, gig bags, beginner books and DVDs, an electronic tuner, picks, and a strap, with free shipping. » *www.banjohut.com*

—

Mandolin Hut

From the people who brought you Banjo Hut, here's the mandolin equivalent. This Knoxville-based mandolin supplier sells complete beginner packages of set-up mandolins and everything you need to learn to play it. » *www.mandolinhut.com*

—

McNally Strumsticks

The ultimate beginner's instrument. » *www.strumstick.com*

—

Morgan Monroe

Great, affordable acoustic instuments. » *www.morganmonroe.com*

—

Jay Buckey

A bluegrass and old-time music archive with plenty of Research, Son for sale, as well as many free PDF downloads of tabbed and noted music. An amazing resource. » *www.jaybuckey.com*

—

Blue Grass Music Jams

Find jams in your state. » *www.bluegrassmusicjams.com*

CLUBS AND ORGANIZATIONS

Sierra Club

World's largest environmental conservation organization. Its biggest perk is that it hosts many local chapters meetings and outings, which you can look up online at its site. Being a member also opens the door to volunteer opportunities, outdoor excursions, and worldwide adventures. » *www.sierraclub.com*

—

Organic Consumers Association

Exactly what it sounds like. » *www.organicconsumers.org*

—

Rural Roots

A Northwest-based community of small farms and organic gardeners dedicated to education and promoting sustainable living. » *www.ruralroots.org.*

FARM FINDING AND SOCIAL NETWORKING

Local Harvest

An online search by ZIP code to find farmers' markets, small-scale farming operations, and organic produce near you.
» *www.localharvest.org*

–

Farmers Only

Want to find some other farmer wannabes out in the ether? Online dating site for farmers, dreamers, and rural-loving people in general. And yes, I'm serious. » *www.farmersonly.com*

–

Sustainable Communities

Take a trip to an eco-village or off-grid community for the afternoon. See how people just like you got into livin' off the land full time.
» *www.sustainable.org*

READING GROUP DISCUSSION GUIDE

DISCUSSION QUESTIONS

1. The word "homesteading" often conjures up an image of a bucolic mountain (or valley) home on several acres of arable land, surrounded by a self-sustaining water system, contented farm animals, and a supportive community of fellow farmers. Yet today's would-be farmer is often a young urban dweller living in a rented home or apartment within a city. Is homesteading all or nothing?

2. Jenna says in her preface, "The work in this book isn't about playing farmer, it's about being more responsible for the tasks we've become numb to." Is it a boon to us that food is available all seasons of the year, or that the television clicks on immediately at the touch of a button?

3. Jenna's journey toward self-sufficiency started when she began researching how products get to consumers. What will the impact of our dwindling fossil fuel resources be on the farm world?

4. Homesteading as a lifestyle can seem "all or nothing." What are some homesteading activities that can be accomplished in an urban environment? What can be done in the short hours between getting home from work and bedtime?

5. How is the burst in popularity of vegetable gardening relatable to our increasingly unstable food supply? Is it possible to entirely control your food supply?

6. Explore the idea of cooperative farming. Can today's urban homesteader make their self-sufficiency dreams come true in a cooperative?

7. While Jenna manages to work a demanding full-time job as a web designer, she acknowledges that her homesteading takes almost all of her nonwork time. Discuss the basic facts of life working in the 9–5 world and maintaining a homestead as well. Is it freedom, or another form of work?

8. A working dog is quite different from a pet. Working dogs need to do their job or their temperament suffers. Can you transform a pet into a working dog? What are your expectations for your animals on a homestead?

9. Jenna describes in beautiful, horrible detail the trauma she suffered when one of her rabbits was injured by her dog and how she had to steel herself to put the rabbit out of its agony. Sometimes farm work is brutally cruel. Is it wrong to give up when the unexpected tragedy happens?

10. Jenna's move from Idaho to Vermont is chronicled in her new chapter "Moving On." In it she explains her decision to move to the East Coast, leaving her friends, in order to make a new life in a new state. Homesteading was a major part of her new life. What changes could you make to embrace a dream?

QUESTIONS FOR JENNA WOGINRICH

What was your first itch to homestead? Were you a "crafty" kid? Did you always want to do this or did it come to you as a young adult?

I wasn't all that "crafty," but I've always been interested in the outdoors and animals. I was the kid who ran off into the woods at Girl Scout camp and was more interested in the reindeer in harness than Santa at the community caroling event. But I fell in love with small-scale farming and homesteading when I lived in East Tennessee. I spent a lot of time in the Smoky Mountains. Part of the national park is a preserved homesteading community called Cade's Cove. It opened my eyes to this way of life.

What is the hardest part of your day? Describe what you do before you go to work, while you're working, and what you do when you come home.

The hardest part is getting up at 4:45 on a rainy Tuesday morning, but it's all downhill from there. I'm happy to do the chores some see as work. To me, time spent outside with the farm and my animals is the fastest, easiest, and most pleasurable part of my day.

Self-sufficiency may seem really overwhelming to most beginners. What is the easiest way to get started?

Slowly and in small measures. Decide to bake a loaf of bread on a lazy Sunday. Or take a knitting, sewing, or beekeeping class at a local college. Get started with friends or your partner, because group activities make it all the more exciting. Getting started could be as simple as a trip to the book store to pick up a book on basic country skills and flipping through it to see what appeals to you. You'll be drawn to certain things over others, and you should follow those instincts. If you're more into animals than carpentry, take a class about keeping chickens at your local extension, or maybe even an introduction to shepherding. Start with good, simple intentions and go from there. No need to raise a barn the same day you decide to start a farm.

You are a vegetarian, but you got a lot of flack when you raised a turkey for your Thanksgiving meal last year. Talk about why you decided to do that (when you weren't going to eat the turkey!) and how you felt about some of the reaction you got.

I'm currently a vegetarian, but don't plan on being one indefinitely. I became one because I didn't agree with the way farm animals are raised and slaughtered — packed into confined feeding operations and processed on an assembly line. So, one Thanksgiving I decided to raise my family's bird, knowing it would have the life I'd feel was morally sound for table birds. To my surprise, some people in my family were appalled at the idea of eating a "pet" turkey. They had no problem serving up a concentration-camp turkey they'd never met, but felt that eating one they'd known personally was murder. It was an eye-opener for me, for certain. When I start raising lambs, things are going to get hairy . . . .

As a result of the book and your blog (www.coldantlerfarm.blogspot.com), you've gotten some pretty major press — a feature story in Bust *magazine, reviews in many major daily newspapers, a gig blogging for both* The Huffington Post *and* Mother Earth News, *as well as a spot on National Public Radio's "Morning Edition." Has it changed the way you go through your day? Are your coworkers intimidated or impressed?*

I don't think my coworkers are intimidated; a few may be impressed (but I doubt it). My office is a small group of tight-knit

Vermonters who just know me as the girl with the orange truck who has a deer head mounted over her desk. Or the girl who brought her goat kid to the office when everyone else brought their dogs. Those who are close friends just see me as a friend, certainly not any kind of celebrity.

Talk about the hard parts — having to put your rabbit down, for example, or giving your baby goat away. Is farming worth the heartache?

Heartache is inevitable, but it is always worth it. As a renter, and a beginner, the experiences are raw and sometimes out of my control. My landlord asked me to get rid of some animals this year and that was rough, but the community from the blog and my neighbors here were so supportive. I feel lucky to have any of these experiences, even the sad ones.

Now let us in on the good parts — what are the best parts of your day, your week?

Any time I'm outside with a hoe, a halter, or a head veil (to take care of the bees), I am happy. Time spent outdoors, even doing menial tasks like moving electric fences or cleaning rabbit cages, fills me with satisfaction and purpose. These tasks produce actual results every time, and it amazes me how rare that is in our modern jobs. At the office I am part of a system. At the farm I *am* the system.

Clearly, a person can homestead wherever he or she lives — you started in Tennessee, moved to Idaho, and then came to Vermont. What do you like best about Vermont? Which is most accommodating to your lifestyle?

Vermont is perfect for me. It's a perfect mixture of all three states I've lived in previously. It has the humid summers, thunderstorms, and lightning bugs of Tennessee. It has the snowy winters of northern Idaho. And it has the feeling of home and comfort of the Northeast I grew up with in Pennsylvania. Vermont's home, and although the short summers aren't ideal for the garden, the sheep seem to do well in the winter, even at –20 degrees!

Talk about your concern about the sustainability of our current food system.

My concerns for the modern system are based around morality and basic safety. To produce that much food, at the price the American consumer demands, seems to invite cruelty and neglect. Every day there is another story about *E. coli* in hamburgers (and sometimes broccoli!?) or a slaughterhouse worker being hurt on the job. Sadly, it seems that the factory system is here until a lot of people get sick at once. So I feel it's our job as conscious consumers to support healthy, local, food systems and to do as much as we can to feed ourselves, or at least support those who are farming humanely. I don't want 2 million people to come down with salmonella before the industry realizes that faster and cheaper doesn't equal better.

What is your one-year goal? Your five-year goal?

My one-year goal at this point (early 2010), is to either buy my own small farm or be well on my way to doing so. I've never owned a home before, so it's been a ride of confusion, realtors, banks, and research. With any luck, fate will land me in the right place for the right price, and I'll finally have my own land, hopefully here in southern Vermont.

My five-year goal is to be writing and farming as much as possible, as well as selling lamb and wool from Cold Antler. Becoming a shepherd with a staff of working border collies is the only thing I truly want to accomplish in the next half decade. It's an uphill crawl, but every month I feel a little closer to reaching my goal.

Can you give readers your best advice for attaining self-sufficiency? What is your kernel of wisdom for a beginner?

The best advice I can give is to start small, and not let those small changes in your lifestyle *feel* small. Changing your life takes time, research, energy, and effort. But changing your light bulbs, grocery bag contents, and library books are great ways to get started.